Globalization, India and the World

By the Same Author
Land Reforms in States and Union Territories of India

About the Authors

Dr. P.K. Agrawal awarded doctorate in Law is a renowned writer of Hindi Literature having published numerous books in English on subjects like Land Reforms, Judicial Reforms, Environment, Pollution Control, Sustainable Development, Indian Bureaucracy, Motivation and Indology etc.

He served Government of West Bengal and Union Government for 34 years as a member of Indian Administrative Service. Presently he is working as partner with Khaitan & Co.

Dr. R.K. Bhatt did D.Phil from University of Allahabad and is Associate Professor of Economics at Banaras Hindu University. His field of specialisation is International Economics and Public Finance.

He has written two books and published various articles as well as research papers. He has been honoured with prestigious Rashtriya Gaurva Award.

Globalization, India and the World

By

P.K. Agrawal and R.K. Bhatt

CONCEPT PUBLISHING COMPANY PVT. LTD.,
NEW DELHI-110059

ISBN-13: 978-81-8069-784-5

First Published 2011

Published and Printed by

Concept Publishing Company Pvt. Ltd.
Regd. Office:
A/15-16, Commercial Block, Mohan Garden,
New Delhi-110059 (India)
Phones : 25351460, 25351794, *Fax* : 091-11-25357109
Email : publishing@conceptpub.com
Website: www.conceptpub.com

Editorial Office:
H-13, Bali Nagar, New Delhi-110 015, India.

Cataloging in Publication Data--*Courtesy:* D.K. Agencies (P) Ltd. <docinfo@dkagencies.com>

Agravāla, Pramoda Kumāra, 1950-
Globalization, India and the world / by P.K. Agrawal and R.K. Bhatt.
p. cm.
Includes bibliographical references (p.) and index.
ISBN 9788180697845

1. Globalization--India. 2. Globalization. 3. Globalization--Economic aspects--India. 4. Globalization--Economic aspects. 5. India--Economic conditions--1947- I. Bhatt, Rajiv Kumar, joint author. II. Title.

DDC 337 22

Preface

Globalization is a reality now. Sooner one accepts it, better it is. It may tumble down, but the niche it has created for itself in all fields will never disappear and will reappear time and again where inefficiency, inaction and delay will bring the system to its brink.

Globalization has shown a dream to the people of the World who were fed up of listening false promises of their political masters and the executives of policies of the polity in all parts of the World. Though globalization also promises welfare of all but its methodology is based on free market and competition. The socialists, dictators and the democratic regimes promised the same in the past but through the process, adopted by them, they could never reach near the goal. The citizens of the present day World are in hurry. They want quick results. They want best of goods and services, facilities, luxuries and entertainment available in every part of the globe. Thus, consumerism of the modern age is the main driving force.

The pilots of globalization mainly the Multinational Corporations or Transnational Corporation or large companies in a country have come forward with a solution that they will be able to deliver goods provided they are given freedom to act and autonomy to decide across the World. World Trade Organization, World Bank and International Monetary Fund have supported them as their mentors backed by the developed countries. The political leaders and their helpers also acquiesced in because everything was being served to them on a platter without any action or hassle on their part.

But the politicians should behave like the statesmen to take into account the destiny of the future generation and the globe and not only their success at the hustings in a period from three years to five years. Their inefficiency, indecision and inaction have brought the World to this brink. This is their solemn duty to take out the World from this morass and give a better World to the future generation to live in. Even the intellectuals have become politicized and start to bend to the ideas of those in power.

It is a fact that globalization is not working to the satisfaction for many of the World's poor as much as for the environment. The transition from communism to a market economy has been badly managed with the exception of China, Vietnam, few Eastern European countries. The gulf between the poor and rich has widened. The regional disparities have increased. There is clear division between the north and south and east and west parts of the globe. The globalization wants to take pedestrians by air. Reform, to be successful should keep pace with culture, thinking and stage of development of stakeholders, it wants to reform so that they may absorb reform and make it their own. The underdeveloped countries are no longer in reckoning of globalization. World Bank has forgotten its duty to look after them. Similarly, IMF is more inclined towards developed countries rather than helping currencies of developing countries which is their primary duty. WTO is openly coming out as spokesman of the developed countries which is against the very mandate of UNO and its bodies. The poorest of the poor in the World should be focus of all international policies and he should be given minimum food and support security so that he can live on his own without being subject to market mechanism. On the other hand, the policies of developed countries are heavily biased towards their national interests and to protect and flourish their economies. It was decided that farm subsidies will be reduced but these were doubled by the USA after the round of discussion at Doha. However, in London discussion of G-20 countries including India, in August 2009, it was decided that USA will open up its economy for export by the developed countries but God knows how far, USA would listen to such multilateral decisions. The multilaterism is just losing the sanctity gradually. This has an inherent risk. If the decision is not implemented by a developing and militarily weak country, the powerful and developed country can go for military solution which will be the most unfortunate end of globalization because neither the UNO nor its Security Council will be able to protect the defaulter developing country. Therefore, multilaterism should be promoted but no binding treaty or protocol should be there. Bilateralism is the well-tried way of dealing between the two countries which should also make a base for globalization. Then the nation-state will get its primary importance because that is nearer to the people and is accountable to the public directly.

World Bank, IMF and WTO should play their role more closely and painstakingly by enforcing various regulations through Governmental, Semi-Governmental, autonomous or regulatory, authorities to be constituted for the purpose. Globalization in ultimate analysis should enhance locations of industries and service centres and should create more permanent employment and better livelihood.

The MNCs, TNCs or national corporations or companies shouldn't be allowed to indulge in flight of capital, speculation or money laundering in the FDIs. The international institutions should step when there is a crisis. The crisis may be of the national level like that of Thailand in 2008 when the Thai Prime Minister was stoned and the Bangkok International Airport was blockaded or may be of regional type like the East Asian countries in 1997 or even collapses of MNCs like M/s. Lehman Brothers, General Motors, City Corps in USA in late 2008 or banks in England or in other countries of Europe in 2009.

An injustice anywhere is injustice everywhere. Globalization should not be another opium for masses rather it should control, regulate or lessen strife or violence like ethnic or religious violence in Latin America, Civil strife in Africa, religious violence in Palestine, Iraq, Afghanistan, North West Province of Pakistan or mafia capitalism in Russia.

In this work, I have been greatly influenced by writings of Joseph Stiglitz, who has so vocally exposed all the facets of globalization before the World community. Various issues of the *Economic and Political Weekly* have been quite helpful. I have tried to build-up by own ideology through articles on 'Work', 'Political Economy of Globalization' and 'Imbalance in Society' because without definite ideological background, writing is just like a ship without a rudder. I have also placed positive aspects of globalization which few writers do and have taken the globalization to lower level agents who can really implement like the Panchaytai Raj Institutions in India and legislators. Modern manager has to be a global manager. I have provided him tips through the topic "Global Leadership". Similarly, the cause of the most down-trodden rural poor has been taken into account in the article on land reforms. Co-author Dr. Rajiv Kumar Bhatt has contributed research based articles on Globalization: Emerging Issues and Challenges for Developing Economies, Impact of Global Financial Crisis on Indian Economy, Role of Service Sector, Second Green Revolution is must to sustain Economy, Inflation: A Challenge For Indian Agriculture and the Role of Micro Finance in the Development of Rural Entrepreneurship. He has also focused on the changing Economic Milieu of India and the relationship between Economic Growth and National Integration. In the end, his articles on India's trade relations with 'United Arab Emirates' and 'China' have guided the trends that the developing countries like India should look towards new markets.

In the end, we thank the Publisher and hope that this book will prove immensely helpful for policy-makers, administrators, economists, researchers and public at large in all countries.

P. K. Agrawal
R.K. Bhatt

Contents

1

What is Globalization ?

Globalization has not yet found the place in the dictionary though it has influenced the present World most. "Globalization refers to a World in which societies, cultures, politics and economies have in some sense, come closer together. It can be defined as a set up of economic, social, technological, political and cultural structures and processes arising from the changing characters of the production, consumption and trade of goods and assets that comprise the base of the international political economy. It is an intensification of Worldwide social relations, which link distant localities in such a way that local happenings are influenced by events occurring many miles away and *vice versa*."[1]

According to Nobel Laureate Joseph Stiglitz :

> "Globalization is the removal of barriers to free trade and the closer integration of national economies. It can be a force for good and that it has the potential to enrich everyone in the World, particularly the poor."

Liberalization is its own brother. It stands for minimal Government or the virtual withdrawal of the state from economic activity. It is not all that bad in principle. It also aims to democratize the system, full freedom to the consumer and reaching prosperity to every home by distribution of excess wealth so created due to process of globalization. Marxism also proclaims to wither away the state. Thus, globalization is competition, quality, cost reduction, innovations and commitment. If one participates with determination in it, it will be highly rewarding for him like China has cornered all the advantages out of globalization and is able to preserve due to its high military might and manpower. The international markets are more cost conscious than the local ones.

It is not big companies which will overtake the small ones but the fast ones which will displace slow moving organizations. Harmony with markets, with all functional areas within the company in tune with the changing requirements of the global market place is a must.

The globalization process has been triggered, encouraged or closely followed by Transnational Corporations (TNCs) or Multi-National Corporations (MNCs) and their associated Foreign Direct Investment (FDI) activities. The concern is how to get maximum rates of return on investment and how it can best be stabilized in long run. Capital always likes to rest in peace and tranquility because her husband Vishnu as per the Indian mythology, is relaxing under sea water.

Globalization is not working for many of the World's poor as much as for the environment. But globalization cannot be abandoned. Neither it is feasible nor desirable. It is a necessary evil. Therefore, we shall have to learn how to manage globalization and to get best returns out of the international situation, we are in. IMF, World Bank and WTO have set the rules of the game. They are always working under pressure from the more industrialized countries. They are not able to remove imbalances created even by so many development rounds of talks like Doha (2001), Qatar, Davos, Delhi, Mumbai, London etc. People oppose the faceless bureaucrats of globalization sitting in WTO, IMF or World Bank yet they do not yield fearing fierce reaction from the developed countries who may stop even aid or their share to run these international organizations where the policy makers are the highest paid employees internationally. The first worry of any bureaucrat is to safeguard his salary and job.

Anti-globalization activists argue that these institutions failed in past because they imposed a wrong development model and in part because they are simply too large to respond effectively to the needs of people on the ground at micro level. They have been promoting a development model that places a premium on maximum trade and investment. This model has been devastating for workers, the environment, equity and financial stability.

Let us examine for the sake of simplicity how different groups, persons or organizations view globalization in India.

1. Social Responsibility of Globalization

Corporate responsibility is of voluntary nature. The current key issues relate to human rights, environment and workers' rights. The companies are socially responsible for both the direct and indirect negative effects imparted upon the respective stakeholders through their operations. The negative effects can take any form including unemployment, in scaling up pollution, non-transfer of technology and exploitation of local raw material. MNCs or TNCs were supposed to take the responsibility to undo these negative effects but they have miserably failed. They are so much busy in profit-making that they neglect these areas. When their profits dwindle, the question to embark upon these non-profit-making areas do not arise at all. The responsibility for the same is left to the inefficient Governments who are already overburdened with their day-to-day activities including production, capital and market management, law and order, public interest and internal political bickering that they cannot turn even to these issues. Even leading industrial houses like Tatas have reduced their social welfare activities because those were expanding exponentially whereas the plant-production was expanding in arithmetical progression. The administrative structure for social activities was becoming unmanageable and was consuming most of the time. For example, TISCO in Tatanagar has its own supply of power, drinking water and are managing municipal amenities for the citizens apart from their own workers and officers' colonies. They are also providing health and medical facilities, housing, cultural activities games and sports in the hinterland of their steel plant at TISCO. As a result, the populace in Tatanagar worship Tatas and no union or association or criminal or mafia can afford to disturb their plant. Their plant is just like the mother for the entire population of about 2 million around. Fearing closure of the plant due to financial burden, Tatas have considerably withdrawn from their social involvement. So also Infosys is slowly reducing their involvement in social welfare activities. However, the familial groups in India and USA in the past had set up large religious, educational, health and charitable institutions respectively but their successors are no longer interested in these activities. In the globalization, everyone likes to be reckoned among ten (10) richest persons or groups in the World or five hundred fortune companies. Their psychological satisfaction has overtaken their sense of devotion to the teeming millions. Thus, globalization has even put the social responsibility in back gear except that they have now started contributing multi-billion dollars to undo emissions of carbon on ozone effect. That is not enough. There is no proper monitoring or social audit system. The civil society or NGOs have been crying horse against them but their voice is not making decisional impact as because the NGOs themselves are not above suspicion. The public is thus becoming restless and the dissatisfied public is a potential danger for bursting the bubble of revolt or bloody revolutions or acts of destruction or arson. The heads of states also gather annually to address the environmental issues since the 1992 Earth Summit in Rio de Janeiro yet the temperature of earth is rising. The ice cap on the two poles of our globe is likely to melt more. Even ice clad Himalayas is not left untouched and the Ganges glacier is shrinking gradually threatening the foodgrains bowl of India. The task is for both MNCs and National Governments to devise ways in which the transfer of environmentally and managerial practices and clean technologies in domestic industry can be encouraged. Technology is the panacea for removing of all ills in the society specially in the era of globalization and the drivers of globalization should effectively prove it and remove poverty from the planet earth so that tears from eyes of the poor do not fall and make this earth cold like moon.

2. From Workers' Point of View

Workers across the globe find that their working conditions have not improved. Nike and Reebok have abusive working conditions in foreign plants most particularly in Asia, which produce for them under subcontracting arrangement. Similarly, garment giants like Levi's and others have factories in India and other countries where people work in subhuman conditions. The factories of MNCs located in Indonesia, Philippines, Malaysia, Taiwan, Thailand do not have those conditions of work which factories in Japan and USA have. In fact, the conditions of workers in Indonesia and Thailand is pretty bad. Intel and Hewett-Packard profess to have their workshops in foreign countries free of pollution but that is not the end of story. IBM does some skill development programme for the local people but the other two giants do not involve in the same. Nike has started to raise standards not just within their company, but in countries where they work through better education and care of workers. Everyone know that the condition of workers is far from satisfactory in China and Russia due to globalization. This is high time that in the race of globalization, the workers' interests and rights are not sacrificed. The workers should also work sincerely during working hours with increased skill. At the same time, they should be paid wages which are not discriminatory and are not inequitable as compared to the management. For a factory, both are equally important. If globalization cannot satisfy its own army of workers, it can never please distant stakeholders like public except through high sounding advertisements, propaganda or rhetoric which will soon evaporate in the heat of people's protests.

3. From Farmers' Point of View

Farmers are happy so far they start getting more remunerative prices of their farm products. As a result of globalization, the demand of farm products increases tremendously as consumption of all types of goods increases as a result of steep rise in purchasing power in general. It is expected that there will be export of foodgrains, fruits, vegetables, meat and dairy products from the undeveloped and developing countries at higher prices through a cold storages chain which will be the infrastructural gift of the globalization. But the developed countries are not fools. They will patronize their agriculture as much of other countries. Everywhere food and thereby agriculture is the base of life. They, therefore, first provide lot of subsidy to their agriculture and if they fail, they impose higher import duty known as dumping duties on imports of foodgrains indiscriminately. As a result, the imports of agricultural products from developing countries cannot enter into the developed countries. This becomes the first set-back to the majority of farmers from the developing countries in the World.

Second problem arises when the large corporations enter into agriculture field in various guises like contract farming, food processing units, chain of cold storages etc. It is a fact that the farmers of the Less Developed Countries (LDCs) are not well educated and they do not know the art of marketing. Therefore, they have to be protected more by the state apparatus.

Third problem arises when there is land acquisition for location of various industries without proper rehabilitation and resettlement policy. So far 67 multi-products SEZs have been started on 1,34,000 hectares of land in India which was acquired mostly by state industrial corporations. The member of the family of the farmer who gets the job in the factory, gets himself alienated from the joint family and family is rendered penniless after few years without any avocation after they eat up all the compensation received in the land deal. It is in this perspective that the land acquisition is being violently opposed by the farmers all over India and even abroad. India is not able to manage Narmada Bachao Andolan, Nandigram and Singur cases of land acquisition. China can suppress the rights of the ordinary people which India cannot do. Thus, the fate of farmers is to be safeguarded in globalization otherwise the farmers like workers of Marx will overthrow popular Governments through violent agitations and army tanks on Tiananmen Square will not be able to silence their revolt.

4. Medium and Small Shopkeepers Suffer

The local shopkeeper used to have close link with the customers. In due course of time, the shopkeeper became more greedy and his subsequent generations forgot the contribution of customers in their family prosperity. As a result, they started supplying substandard goods and services and misbehaved with the customers or the clients. This alienated the customers from him. The customers found an alternative in sprawling shopping Malls, glittering stores and restaurants and multiple shopping centres. They also provided one window shopping at a cheaper rate even cheaper than the local shopkeeper but not cheaper than the small retailer, tiny, roadside shopkeeper or hawker who has to spend nothing on his establishment. The buyer or service beneficiary, therefore, finally, turned to the newly created shopping Malls run by the MNCs and large national corporations, companies or entrepreneurs. This has caused unrest among the Indian middle class which has its majority in middle class shopkeepers or service providers. He is the worst sufferer of globalization. The consumer at present is being benefited and is enjoying his enviable position. Though these shopping Malls employ large number of employees yet that is not enough. In a middle class shop, his whole family is employed. What else will that unemployed person do? Now he is the front ranker opponent of globalization. But he should not forget that globalization has flourished due to his deficiencies in the past. If he corrects his deficiencies, his customers will again go to him as he has traditional linkage and emotional bond with him which he can never have with these large and indifferent monsters of globalization. In the long run, globalization is also going to die its own death due to inefficiency and high cost administration which is in-built in such organizations if not properly controlled. A bottom heavy person cannot move due to his weight.

5. Attack on Culture

According to Fiedel Castro of Cuba, various cultures in return will attack the globalization and will finish the demand of globalization. Literature is the base and offshoot of culture. Globalization proposes to have uniform culture around the world because this helps in its aim to bring uniformity so that all people think alike, act alike and speak alike. This will facilitate for marketisation and commercialisation if the relations among various groups of society and nations can be patterned uniformly. It will help the MNCs to produce products in country of their choice at minimum cost and will sell in those countries where those will fetch maximum profit. To accomplish this, the alternative globalization will move beyond economic growth and trade driven development. India will be benefited most because its culture will have the highest export value after its software exports and accounting experts (CAs). But it is neither feasible nor advisable. It will not happen. Asian, African and Latin American cultures and cross-cultures will revolt against. On the other hand, globalization will never be able to control corruption nor it will be able to provide better governance because before it dictates in formation of Governments, it will die its natural death. For example in between India and Pakistan, Kashmir is the most important issue and will remain more important than the trade. Similarly, between India and China, border issues are more significant than trade. India can never accept any part of Arunachal Pradesh as part of China. Thus, regional and local issues will be more important and thereby regional and local cultures are more easily unified than the global ones as they have been easily interacting with ease for many centuries. For example, in spite of rush of billions of dollars in Pakistan on the behest of political and army leaders of Pakistan, the people of Pakistan are treating USA as their enemy number one. So is the story of many other Islamic countries like Iraq, Iran, Afghanistan, Syria, Libya, Lebanon etc. In spite of effect of Marx for more than a century, the religion could not wither away. So also the Governments will not wither away so also cultures. It is high time that globalization realizes the strength of various cultures and joins them with a thread of peaceful co-existence so that people may enjoy the variety of their cultures along with the economic variety and prosperity brought about as a result of globalization.

6. From Media Point of View

Media is the happiest class in the globalization. They have not to produce various programmes for different people in hundred and thousand varieties in different languages or background. As a result, they earn huge profits. The media barons, therefore, are the strongest supporters of globalization. Through computer and wave, they will have easy reach to any part and person of the planet earth where even political, religious or social leader cannot have access. They will become the top pressure group and will think that they may make or unmake a Government. The newspaper, magazine or bestteller books in USA and UK will be read and be followed all over the World and franchise business will suffice. This will happen at the cost of culture and souce of variety. Instead of encouraging morality and honesty, the media will indulge in spoiling of the minds of young generations and the working people alike. However, none will listen to the old people who will continue to criticize the globalization and will be the most active though silent supporters of their dying cultures. Like middle class shopkeepers, old men and women will dislike the media run by the globalization and will be the worst sufferers. There will be only lip service to them. Similarly, the farmers will lose focus in media manoeuvres. Only groups who are important to the decisions of purchase of commodities and services will find place in media to earn huge revenue from advertisements or propaganda. Later, many of such alienated media companies from masses will be wound up.

7. Judiciary and Legal Profession

These two classes will be the choice beneficiaries of the globalization. Because of high cost of litigation, only MNCs or large national corporations can pursue litigations successfully and pay high costs to solicitors, advocates, lawyers or attorneys. Ultimately all local disputes will end up in appeal and multiple appeals. Therefore, the expert in legal professions will have skyrocketting income whereas normal advocate at the grassroots level will suffer. The Judges in higher judiciary will be the new icons of the society though they may not have direct touch with the ground realities in the field. Law will be what lawyers will dictate and judges will accordingly pronounce. Even they will jointly control the day-to-day affairs through public interest litigation. Delay in delivery of judgments will be part of judicial process. Everyone in the judicial system will relish being a part of the system even without any outcome from the judicial process. Thus, the globalization will have its vocal supporters in legal professions because it is very easy to talk about eternal and universal principles of good polity, i.e. secularism, demoralization, equality, liberty, individual freedom, freedom of press, human rights, environment and pollution control. None will have time to talk about starvation. Incompetent bureaucrats and corrupt politicians will not have moral courage to oppose increasing interference in the management of polity by the judiciary. It is a different story that corruption will eat into the vitals of judiciary too.

8. What Bureaucracy Thinks of Globalization

Bureaucracy in every country will be as usual slow in reacting to globalization because globalization basically challenges their *status quo*. They will find in ultimate analysis that they will have no alternative except to adopt globalization because it has benefited so many similar other countries. Secondly, globalization will bring in millions of extra dollars to their exchequer which they will be able to spend or utilize easily without much efforts. They have simply to print the national currency which is the easiest job for Government without creation of industrial, communication or social infrastructures which are supposed to be created by the MNCs, TNCs or their national counterparts in the garb of joint ventures. Thus, they will find the easy wayout to manage the Government works which they gradually find it difficult to feed the white elephant of the bureaucracy. Their political masters will ultimately give in because their voters also like to enjoy foreign goods and services. They can no longer pursue or advocate slogans for indigent to the masses without making their own system efficient and transparent. The bureaucracy and the political elite have ultimately to give in to the globalization.

9. From Women's Point of View

Women are another group which is the sufferer during the globalization. Though the drivers of globalization profess gender equality but that is only a lip service. Some educated and professional women get top jobs in the MNCs and companies, yet this is only the tip of the iceberg.

Young women also get highly educated and are happy. However, the major constituency of women folk, i.e. the domestic women suffer. Their husbands spend most of their time on entertainment and in luxuries. Everybody is in front of television. Everyone in the family is lost in a dream world of luxury and ostentatious living on the pattern of USA, European countries, Canada, Switzerland or Australia yet none realizes the ground realities of life which are quite different. The prices of essential commodities keep on rising. The labour becomes dearer. No domestic help is available. The modern equipments for domestic chores like washing machines, microwaves, micro ovens, water filters, vacuum cleaners, utensil cleaners and other kitchen equipments become very costly. Their maintenance becomes constant headache for them. The maintenance becomes so costly that their replacements become cheaper. Thus, there is constant need for goods and services which their wage earners cannot afford. More women have to resort to prostitution because that job becomes in high demand and well paid. The male folk lose attraction towards them as everybody in the society is running for something which he does not know himself. They adopt some fashion at the instance of their male folk against their traditional mentality which creates psychologically disorder in them. The women labourers are another class which is sufferer. The women though projected to get equal wages with man, in fact they do not get the same. They do not have vacations or intervals to look after their children. Though new devices like crèches come in the existence but in practice, those devices do not serve full purpose as all solutions of the globalization are indifferent and mechanical. When women become more than males in jobs, the husbands become mentally sick at home.

10. From Youngmen's Point of View

The young boys and girls rally in the environment of globalization. Everyday they have thrills in their life style. They have new devices to communicate and to get educated. Their mobile sets get so sophisticated that only one device can fulfil their all requirements. They have i-pod, computer, laptops and what not? They communicate globally and there is speed, fast life style, fast food which they like. They have global dreams and in fact they move globally. They earn so much by working for five-six years which their predecessors did not earn in their lifetime. The greatest advantages that their talent is respected and there is no obstacle if they can have money. They get high salaries in start-up jobs which was never there but they forget that the salaries will not rise much subsequently. There is no stability or permanence in their jobs yet they do not mind because job becomes available elsewhere without any waste of time. They can enjoy best of food, best of cars and bikes and can have best of garments and posh life style. They also find smooth working atmosphere for them. Tradition or convention does not stand in their way. As global managers, they can draw higher salaries than the richest man earning in their village or urban neighbourhood. They can remit lot of money to their parents, countries and can have lot of savings after lot of fun. They have high mobility which they like. However, they forget that this will speak adversely on their physique subsequently. But they became self-assured that the high rise sophisticated hospitals as modern temples, will look after them because by that time, they will have lot of personal savings in bank accounts.

Conclusions

In nutshell, it is difficult to raise theoretical war against globalization at present. Acquiescence is the best solution. Globalization has in-built mechanism to correct its pitfalls and that too quickly. Therefore, globalization and liberalization need to be given trial. Like any other system, globalization has also some major shortcomings. It is hoped that globalization will bring in better livelihood for all the people and the downtrodden section of the globe will usher in new era of development. To be called successful, globalization must reduce poverty and must remove the black spot of starvation from the face of the planet earth. It is said that mother earth has enough resources in its womb to feed her children. This should be proved in reality but keeping in mind that it does not result into degradation of forests, rivers, mountains or other natural resources because that will be very fatal for the future generation in the long run.

There is a challenge before globalization that it makes the World a better place for living and we leave the earth in a better shape for the future generations than the one we inherited at our arrival on this earth.

NOTE

1. Bende Anthony, Nabende: *Globalisation, FDI, Regional Integration and Sustainable Development*, Ashgate Publishing Company, Burlington, USA, p. 7.

2

Political Economy of Globalization

In the past, the state is considered as godlike benefactor aiming for social welfare. It was accepted that king can do no wrong. As a corollary, the king had the power of 'eminent domain'. He can acquire land etc. and take any other action in the public interest.

Adam Smith proponented that there should be an intercourse of trade and commerce. Free market will build-up prosperity through competition and development of skills.

Karl Marx exploded all the previous theories in a very scientific way based on historical facts and analysis of economic data. His arguments still stand with full bloom. However, retreat began with Downs (1957). He argued that voters and their representatives maximize their utility functions in the political process not less than in the market and inferred from this that competitive political markets would ensure pareto-optimality of the policies pursued under majority rule.

In nineteen eighties, a few groups of the eminent Economists like Bhagwati and Srinivasan (1980) sought to model the processes that lead to protection or liberalization. The protection policies help the Government in power to stay in power and not to allow the opposition to grow. The public get confused as to what the causes of failures are: whether these are internal or external or depend on weather and is unable to monitor the performance of the Government. This gives advantage to large established or concentrated industries over the small scale heterogeneous ones. The consumers are not organized into an effective pressure group. Therefore, their impact on commercial policy will be minimum. Thus, MNCs flourish. A new intermediary group is created who gets discretion exercised on selective individuals or companies and gets vested interest in permit and licence raj (regime). The liberalization professes to do away with this group yet such a group flourishes more when liberalization becomes bottom heavy, i.e. more plan expenditure rises sharply. The labour and capital have similar support though their long term interests may diverge sharply. Thus, liberalization goes on and on.

The development and democracy are must in today's World. The state, therefore, does a good job of transferring resources from the accumulating economy of corporate capital to the dispossessed masses, thereby reversing the *effect* of primitive accumulation otherwise the political party will lose at the election and thereby the power and authority.[1]

The degeneration and corruption of the Marxist socialist system, leading to the collapse of the Soviet Union, and the cooption of China into the global capitalist system, have in a certain measure led to the devaluation of the ideology of socialism as a superior alternative for a large number of people around the World. Some among them have been led by political charlatans to seek alternative models of socio-economic transformation possible in religious utopias, to reach which they have been inspired to traverse the dark alleys of religious terrorism. The present Indian situation is no exception to this trend.

It is in such a global situation – a political vacuum created by the collapse of the socialist system, and an ideological void waiting to be filled with fresh theoretical concepts for restructuring a new socialist society – that neoliberal capitalism has emerged as the decisive force in a unipolar global system under US hegemony. As in the last days of Engels at the end of the 19th century, capitalism today has been able to harness powerful technology to expand its tentacles. It concentrates on two crucial sectors – information and military. The technological sophistication in both areas has rendered the tools used by the opponents of capitalism today – in the words of Engels – "to a considerable extent obsolete." His followers today will have to invent new tools – both politically and logistically functional – to match and defeat their opponents.[2]

It is now widely agreed that the speed of the reforms in the former Soviet bloc countries was a mistake. The privatizations were done before sound regulations and strong tax laws were put into place.

As Government revenue dropped, spending on health and infrastructure collapsed. One of the legacies of Russia's past was a high quality education system, but this quickly deteriorated as Budgets were slashed. At the same time, the old social safety nets were being cast aside. The results were grim poverty in the former Soviet bloc countries increased from 1987 to 2001 by a factor of ten.

But story of China and Vietnam was different. They while retaining a Communist political regime, also began to move to a market economy, and the contrast was striking. They rejected shock therapy in favour of slower and more gentle transition to a market economy. Today the vibrancy of their economies suggests that one should not give his identity or ideology. In case of European Soviet bloc, the countries like Poland and Slovenia managed transition better because they became members of the European Union.

Economics is the science of choice. Bangladesh's micro credit rural banks have done wonders for mobilizing women in conservative Islamic Bangladesh. Village irrigation project opened doors of prosperity in Nepal. Schools in Bhutan are doing wonder in remote villages to change the phase of Bhutan which is closed to the World. In India, growth of self-help groups specially of women is going to change India. In mountainous parts of China, hard-work by its people to manufacture handicraft goods and mini steel-plants have revolutionzed China's economy. Similarly, Islamic guerillas in Mindanai (Philippines), Afghanistan, Iraq and Pakistan are bent upon to destroy all reforms in the economy taking alibi of globalization as creation of America and decrying strongly the role of USA as the World policeman. On the other hand, in the 1990s, many of the African and Latin American countries, including Nigeria, Kenya, Tanzania, Uganda, Ethiopia, Ghana, Brazil, Mexico found problems with their leaderships. The political leadership could no longer create euphoria on some emotional, cultural or regional issue and get away with it even after indulging in corruption and providing an inefficient administration to the people. The people wanted better life like Americans and they were not prepared to listen to false promises and baseless advices from their leaders like Idi Ameen in Uganda and Gaddafi in Libya, Saddam Hossain in Iraq and Khumini in Iran. As a result, new leadership grew up. The new leaders seemed more committed than the old ones to pursue good economic policies. Deficits and inflation were brought under control. Some, such as Olusejun Obasanyo in Nigeria, Yoweri Maseveri in Uganda, Benjamin Mkape in Tanzania and Meles Zenawi in Ethiopia took strong action against corruption. Ethiopia grew at more than 6 per cent annually between 1993-1997, when war broke out with neighbouring Eritrea. Uganda grew, an average, more than 4 per cent annually from 1993 to 2000.

In the above backdrop, let us examine how the political economy grew in India after independence in 1947 which is equally relevant for the World also. The Gandhian ideology is more relevant to the present day World than ever before. Gandhiji said that wealth in the village is equal to the production minus outgo from the village, i.e.

$$W = P - E$$

It does not mean that Gandhi was very conservative or he did not advocate for export. He meant that, whatever way the market forces may work, overall wealth with the villages should be sufficient to provide them two meals a day. The rural people cannot negotiate with the mechanism of fast, clever and manipulative market. Therefore, at any time, he should have sufficient food stock to eat, clothes to cover his body and roof to protect him from vagaries of nature like rains etc. Therefore, he had permitted export from the village which will keep economic balance in the village:

> "There is no natural right in property and as Burke said in his Reflections, 'the Government is not made by virtue of natural rights, which may and do exist in total independence of it'. Natural rights embrace activity outside the status of citizen. Legal rights are required for free existence as a social being and the State undertakes to protect them. Fundamental Rights are those rights, which the State enforces against itself. Looking at the matter briefly but historically, it may be said that the Greeks were not aware of these distinctions for as Gierke points out they did not distinguish between personality as a citizen and personality as a human being. For them, the individual was merged in the citizen and the citizen in the State. There was personal liberty and private law but there was no sharp division between the different kinds of laws. The Romans evolved this gradually, not when the Roman Republic existed, but when the notion of a Fiscus developed in the Empire and the legal personality of the individual was separated from his membership of the State. It was then that the State began to recognize the rights of the Individual in

> his dealings with the State. It was Cicero who was the first to declare that the primary duty of the Governor of a State was to secure to each individual in the possession of his property. Here we may see a recognition of the ownership of property as a Fundamental Right. This idea was so engrained in early social philosophy that we find Locke opining in his 'Civil Government' (Ch.7) that the Government had no other end but the preservation of property. The concepts of liberty, equality and religious freedom were well-known. To them was added the concept of property rights. Later the list included 'equalitas, libertas ius securitatis, ius defensionis and ius puniendi.' The concept of property right gained further support from Bentham, Spencer, Kant and Hegel. The term property in its pristine meaning embraced only land but it soon came to mean much more." [3]

India experimented with mixed economy, K-industries, decentralized planning and things like that so that welfare may percolate down to people. But required results were not achieved.

In the meantime in nineteen eighties it was noticed that the capital started moving globally due to information technology. The world became a global village. It is most striking feature as the run away finality of global finance, which appears remarkably independent of traditional constraints of formation, transfer, national regulation, industrial productivity or 'real' wealth in any particular society, country or region. The predatory capital finds support in common consumer who gets better quality of goods and services at lesser prices which so many earlier votaries of democratic decentralization could not provide. The public at large was no longer interested in baseless slogans by the political masters. They want goods and services at the cheapest prices comparable with the best in the World. The media has created such a demand in their minds that the public cannot get away from this mirage. As a result globalization flourished finding its support base, for the first time in history among the common man.

China's model of globalization is the ideal one to be adopted by the developing countries. Young party leader Hu Janato in association with Wen Jiabao adopted globalization while emphasizing strongly to root out the corruption. Jiang Zamin said, "If we do not crack down on corruption, the flesh and blood ties between people and the party will suffer a lot and the party will be in danger of ruling position, or possibly leading for self-destruction."[4] National Congress of the Communist Party of China known as CPC resolved to implement three R's as follows in November 2002 :

1. Development trend of China's advanced productive force.
2. Orientation of China's advanced culture.
3. Fundamental interests of the overwhelming majority of the Chinese people.

They accepted that science and technology are the primary production forces. But it has to preserve and improve people's livelihood and in consonance with the Chinese culture and identity.

China encashed on the weakness of USA economy. In USA, being the storehouse of gold and dollar, the people consumed more than 20 per cent of the World's productions but they do not want to work by hand. Variety of goods lay only in handwork, and people in globalization like variety in all fields. Variety is the sauce of life. It suits to the market as there is a constant demand of goods. Market drivers ran a mock to the developing countries to import and manufacture cheaper goods because as a matter of policy, it was allowed to import anything or everything at least by USA multinational corporations. As a result labour intensive articles made of wood, straw and cane including wicker work and baskets, head-gear and umbrellas etc. flooded US markets from China. China motivated its workers to manufacture these goods on cheaper rates by taking lesser wages and working for more hours than prescribed ones in factories. At so many places, it amounted to more exploitation of working class than even in capitalist countries. It was a challenge to make China a rich country and strong nation. India also followed China but could not comprete due to cheap and disciplined labour in China. USA was panic-struck. Their dollars were going into the state exchequer of China and other developing countries. USA thereafter, did not open its markets further. They provided more subsidy to their agricultural products which was against the WTO guidelines. They imposed restrictions on imports on some plea or the other like employment of child labour in textile and carpet sector in India and handicraft sector of China and imposed abnormal dumping duties. European Union followed, as usual in the footsteps of USA. The European countries like Germany, France, UK, Australia and Canada imposed restrictions on imports from China on the plea that products like toys were not safe for the children just because the one or two children swallowed the toys.

It is a natural process. It is seen and Karl Marx has correctly stated that the law makers become in league with the capital apart from the religious heads. In village or town, or society, the head or 'panch' violates norms set under his Presidentship when it comes to apply in his personal matter.

In that case, his personal interest is the uppermost. Similar is the situation in the international arena. The domestic policies and interest dictate international policies or relations. So far these suit the national citizens specially in case of the developed countries, it is alright. The political leadership has the mandate to provide them the best life-style. These leaders sometimes plead openly for their country's MNCs. These countries in the group of G-8 and extended G-8 met many times and there were many widely represented convocations, conclaves and conferences but the diplomats of the developed countries showed their back when the resolution could not craftly drafted in their favour. If it was drafted against them as they are in microscopic minority, they contemptuously went back and drafted domestic regulations which could serve their national interests only. But by this time, China, India, Brazil, Bangladesh, Sri Lanka, Nigeria, had made lot of foreign exchange in their 'reserve' to last for few decades. However, the developed countries cannot remain silent spectators, to the prosperity of the developing countries and they will evolve process of various sanctions. Sanctions generally do not work in international field as every country is sovereign. If he refuses, he can very well remain aloof after earning without opening its markets to foreign financial institutions or to high technology MNCs in tele-communications, media etc. This may invite possibility of war among nations on the issue of capture of markets through overtly reasons shown will be different as violation of trade treaties or to compel compliance of international co-operation and protocols. UNO as usual will remain silent spectator and WTO, World Bank and IMF will sing at the tune of the developed countries because the architects and bosses of these institutions basically believe in market economy and feel that what USA does, is right.

Whatever the case may be, globalization has come to stay. Most important, however, globalization has an instrumental role to play in the development process of all countries. What is required are global, national, regional and local policy instruments that can maximize its potential benefits, yet minimize its potential hazards. Most varieties of Marxist theory, though sharply critical of the capitalist conspiracy behind modernization theory, nevertheless were equally realist both in their picture of the architecture of the World system and their understanding of the relationship between theory and practices.

Therefore, the developing or undeveloped countries should improve investment and environment, promote domestic capabilities, develop integrated strategies, build educational bases, strengthen technological institutions, encourage firms into export markets to test and advance their competitiveness and pursue regionalism based on economic rather than political motives. This framework if properly planned and appropriately implemented, will enhance and sustain their economic growth rates. On the other hand, for globalization to last long, the aforesaid economic benefits should be distributed equitably. Thus, globalization should lead to less environmental destruction, less disparity within and between nations, less marginalization of people and countries, less violation of human rights, less mire ability of societies, less vulnerability of people, and less poverty and deprivation.

India always believes in adopting reforms slowly. Therefore, there is not much danger to the Indian economy from any innovation like globalization. India is a self-sufficient country having strong agriculture and manufacturing base. The globalization as such, cannot be taken for granted unless it is transparent and reduces poverty in the country and helps in increase of equality among her citizens. It is very strange that globalization is a concealed friend or foe. It is invisible but moves so fast that one cannot catch it from the front. One will at the most be able to catch hold of its tail. For example, the Americans could only wake up from their globalization slumber after Laymen Brothers, Enron and few other major corporations went bankrupt in 2008. Then, the American policy-makers realized that it was too much of globalization. They have to increase their working hours on work sites, reduce their expenditure on personal consumption and their children have to work hard in studies otherwise the Indian and Chinese youths will overtake them and make them slaves in their country. This can happen with any country.

The message from the temporary collapse of the part of the financial system of USA is very clear that one cannot afford to be complacent in globalization howsoever big or advanced he may be. One has to constantly endeavour to work efficiently and must win in the competition. The management during globalization era has to be fast moving, constantly innovative, fast decision taker and prompt handler of emergencies. It is but certain that the political leaders in the developing countries like India cannot shoulder the burden alone nor their trusted bureaucratic setup can cope with the challenges of changes brought about by the ever expanding markets. As a result, there will be a partnership between polity and business. The political leadership will ultimately give way to the capitalist business and markets. But the people at the receiving end will not accept this unholy alliance. They will keep on changing the political leadership unless business tycoons are fed up and surrender. One of the offshoots of this phenomenon will be the worst that the domestic capital which has multiplied during the globalization will take the flight to the least developed countries (LDCs) where they can dictate policies or will take refuge in those countries where they can employ their cheaper human capital from the developing countries. The fugitive capital will face revolt in the developed countries too yet this capital specially grown in the mechanism of globalization knows the art to run here and there. Ultimately, it will reappear with more vigour, strength and glamour at the appropriate place and time:

> "A developing country needs to be careful in dealing with these changes as it seeks to open itself to the currents of global business. The overarching architecture of treaty law, dispute resolution, management and legal expertise favour the first World in general and the global corporations in particular. The asymmetry and initial conditions have to be offset by vigorous application and exercise of caution. Some of the social evils and economic deficiencies that evoke raging passions are attributable not to globalization but to inadequacies and deficiencies internal to the polities. These have to be accepted candidly by the political class. It is not free markets that have caused inadequacies in social services, but self-centred interest groups that have used policies designed to deliver social welfare for perpetuating themselves. National rhetoric and ideology have become the rallying point for interest groups that rush to find foreign scapegoats to blame for heir incompetence."

This is a good achievement of the modern economic and political patterns. Now the persons-in-power cannot show alibis for their failures. They have taken the advantage of ignorance, illiteracy and patience of public for a long time. Now time is up. They have to perform or perish. The capital is a fast moving being in today's World like air (*pawan*). It has competition and market as its two feathers. Its mind is consumerism. Its belly is quality. Its two hands are skill and knowledge. Its two feet are information technology and land. Capital is flying. It depends whether a country is capable of catching it and to enslave it. It is possible now-a-days which was not possible when Marx had predicted. The warning of Marx about capital is fully relevant today. If a nation or society cannot control the capital, it will enslave the nature and people. However, there is no escape. One cannot live in isolation in this World. Therefore, none should avoid work at any cost. That is the only asset left with the working class. They should be vigilant to their rights and equally to their duties. They should identify their friends and foes simultaneously. If any leader asks them not to work, they should discard him because he would not come for their assistance during their starvation. They should closely watch the lifestyle of their union or proletariat leaders as well as political leaders of the country. If they only preach and indulge in rhetoric, they should be abandoned. Time is not far when people will join and revolt against political leadership. The political leadership, therefore, should be abundantly cautious. The business should also remember the message of Mahatma Gandhi that they are holding wealth on behalf of the society as trustees. In fact, they are the managers of public wealth and therefore, they have an onerous responsibility to discharge their duties and responsibilities with due care, sincerity and observe austerity. Any abnormal behaviour on their part of ostentatious living or show of wealth will put them in the bracket of enemies of the society.

It is hoped that public will exercise due vigilance and will evolve and manage self-governing local units properly so that they are ruled by themselves. All should fight the enemies together known as corruption and delay. If this happens, the failures of globalization will not adversely affect the downtrodden people. There will be lot of cushion. The success of globalization or for that purpose of any polity lies how it reaches and ensures livelihood to the last man in the society and makes him stand in life with dignity. The lowest of the society have also to safeguard that after marketing, adequate quantum of food or goods are left with them for their proper livelihood. It cannot be left on markets or on their rulers alone. People ultimately have to be self-rulers.

NOTES

1. *EPW*, May 26, 2009, Vol. XLIV, No. 22, pp. 157-58.
2. Banerjee, Samanta: Hobson's Choice for Indian communists, *EPW*, May 7, 2005, p. 1935.
3. Agrawal, P.K. (1993): *Land Reforms in India*, M.D. Publications Pvt. Ltd., New Delhi, p. 41.
4. Mohanty, Manoranjan; CPC's Fourth Generation Ideology, *EPW*, April 2003, pp. 1367-68.

3

Globalization in Historical Perspective in India

The package of economic reforms in the initial years after independence of India was based on the perception of economic realities at that time and the newly adopted constitutional framework.

The share of India's export economy in total World export was than maximum which got reduced subsequently as follows :

Year	*India's share*
1950-51	1.85%
1960-61	1.85%
1990-91	0.5%
2000-01	0.7%
2005-07	1%
2008-09	< 1%

It shows that India has no other option except to open out to the World's economy just to keep herself at 60 years' past level.

India started her journey on four pillars of planned development, mixed economy, regulated economy and self-reliance. There is no surprise if in India, the land of variety and a vast country, one single solution will not work-out fully.

The achievements of the Indian economy with regard to creating a more or less self-sufficient production structure are quite impressive.

Moreover, the Indian economy cannot afford to be in isolation so far requirement of petroleum products is so heavy in the country. It is export which can ensure smooth flow of import irrespective of valuation of the currencies.

Percentage of Imports to Indigenous Supplies

Sl. No.		1950-51	1955-56	1960-61	1965-66	1973-74	1977-78
1.	Food grains	5.9	1.7	4.7	9.5	4.3	0.2
2.	Iron & Steel	25.2	39.9	35.7	16.7	10.5	1.1
3.	Machinery	68.9	41.0	40.7	27.8	17.0	15.3
4.	Petroleum	92.5	93.8	94.6	76.6	70.8	63.1
5.	Nitrogenous fertilizers	72.5	39.8	80.3	58.3	38.3	27.5

During 1973-79

The Indian economy grew at a faster pace during the period 1973-77. The growth was at an average of 4.3 per cent a year. This happened in spite of the fact that oil prices rose dramatically and caused disruption in pattern of trade and growth in the rest of the World by OECD group. It was due to increase in productivity of Indian agriculture and easy foreign assistance. From 1971 onward, India started accumulation of substantial foodgrains as buffer stock after learning a lesson from the drought of 1964, resulting in a famine like condition in the country in the subsequent years. This was also due to shift in export trade to the less developed countries.

The gold prices were also increased so as to make at par with foreign market of gold to reduce smuggling trade.

The Period 1979-83

The second oil-shock in 1979 and the deep recession in the industrialised World began in 1979. As a result, the trade deficit mounted in 1980-81 to a staggering Rs. 5,967 crore. The country did a fantastic job. The Sixth Plan allocated Rs. 2,874 crore for the exploration and production of crude oil which went up from 10.5 million tonnes to 26.3 million tonnes in 1983-84. So also there was a thrust on 90 per cent increase in the domestic production of nitrogenous fertilizers. The programme of electrification of railways was pushed up. India's export earnings continued to grow up by about 10 per cent. Traditional items like tea, jute and cotton textiles continued to lose grounds whereas engineering goods, chemicals, gems and jewellery, handmade carpets and handicrafts did reasonably well even under recessionary condition. IMF loan also eased the balance of payment situation.

However, due to bureaucratic control and slow decision-making, the capacity utilisation and thereby industrial production decreased from 74.5 per cent in 1972 to 69.38 per cent in 1982. But all the steps taken through Five Year Plans made the Indian Economy as highly vulnerable but diversified one supported by a protective trade regime.

The fiscal situation had been under mounting pressure through the 1980s and assumed crisis proportion by the beginning of the fiscal year 1991-92. The inflationary spiral which began in October 1990 and reached its peak in August 1991 at 16.7 per cent (as against the annual rate of inflation of about 7 per cent in August/September 1990), became a cause of serious concern. The first signs of the current payment crisis became evident in the 2nd half of 1990-91 when the Gulf War led to a sharp increase in oil prices. Foreign exchange reserves began to decline from a level of $3.11 billion at the end of August 1990 to $8.96 million on 16th January, 1991.

By the beginning of 1991-92, it had become clear that the payment crisis was no longer due to trade deficit, which had come down from $781 million per month in October-December 1990 to $382 million per month in January-March 1991 and further in April-June 1991.

By June, the balance of payment crisis had become overwhelmingly a crisis of confidence in the Government's ability to manage the balance of payments. A default in payments, for the first time in the history of India, had become a serious possibility in June 1991. Suppliers became reluctant to sell goods and services and insisted payments through the banks of their country. This led to industrial dislocation, shortages, specially of oil and high inflation. The new Chandra Sekhar Government leased 20 tonnes of gold out of its stock to the State Bank of India to enable it to sell the gold abroad with an option to repurchase it at the end of six months. This export took place in May 1991. In addition, the Government allowed Reserve Bank of India to ship 47 tonnes of gold to Bank of England in July 1991. This helped to raise about $600 million. The exchange rate of the rupee was also adjusted in July 1991 to bring it to acceptable levels which could be defended. Later on, all the hypothecated gold was got released and was transported back to the country's reserve. It was evident that the economy needed substantial reforms if the crisis was to be fully overcome. One of the important policy corrections apart from fiscal and trade policy reforms was industrial policy reforms.

Industrial Policy Reforms

In order to consolidate the gains already achieved during the 1980s, and to provide greater competitive stimulus to the domestic industry, a series of reforms were introduced in Industrial Policy. The new Industrial Policy of 24th July 1991 sought substantially to deregulate industry so as to promote the growth of a more efficient and competitive industrial economy. Industrial policy reforms announced in July 1991 should be seen as being complimentary to those undertaken in trade and fiscal policies and in the management of the exchange rate and the financial sector. The central elements of these reforms were as follows :

1. Industrial licensing was abolished for all projects except in 18 industries where strategic or environmental concerns are paramount or where industries produce goods with exceptionally high import content. With this, 80 per cent of industry has been taken out of the licensing framework.
2. The MRTP Act was amended to eliminate the need for prior approval by large companies for capacity expansion or diversification. This will enable Indian firms to become strong enough to compete effectively in global markets.
3. The requirement of phased manufacturing programmes was discontinued for all new projects.
4. Areas reserved for the public sector were narrowed down, and greater participation by private sector was permitted in core and basic industries. In place of the 17 areas earlier reserved for investment by the public sector, only eight such areas are now reserved. These rights are mainly those involving strategic and security concerns.
5. Government clearances for the location of projects was dispensed with except in the case of 23 cities with a population of more than one million.
6. Small-scale enterprises were given the option to offer up to 24 per cent of their shareholding to large scale and other industrial undertakings. This will provide them with greater access to capital and technology.
7. Loan agreements of the financial institutions with privately managed firms were earlier required to provide for the right

of the financial institutions to convert the loans into equity. In August 1991, the institutions were permitted not to insist on this provision in future loans unless they felt it necessary for commercial reasons. In December 1991, the institutions were permitted to delete the provision from past loans also subject to a revision of the interest rate where appropriate.

8. A National Renewal Fund has been set up with a corpus of Rs. 200 crore to ensure that the costs of technological change and modernisation of industry would not be borne by workers. It will be used to provide a safety net to workers in sick and non-viable enterprises, and to finance their retraining and redeployment.

Along with a reform of industrial policies, steps were also taken to facilitate the inflow of direct foreign investment. These non-debt-creating inflows will reduce reliance of fixed-interest debt and also bring in new technology, marketing expertise and modern managerial practices. The following measures were taken in this context :

1. The limit of foreign equity holdings was raised from 40 per cent to 51 per cent in a wide range of priority industries. However, foreign exchange outflow on account of dividends on additional equity will be balanced by export earnings. Such foreign equity participation now has automatic approval and is cleared by Reserve Bank of India.
2. The procedures for investment in non-priority industries have been streamlined. The Foreign Investment Promotion Board has been established to negotiate with large international firms and to expedite the clearances required. The FIPB also considers individual cases involving foreign equity participation over 51 per cent.
3. Technology imports for priority industries are automatically approved for royalty payments up to 5 per cent of domestic sales and 8 per cent of export sales or for lump sum payments of Rs. 1 crore.

Deregulation will reduce the role of Government Regulatory Agencies. They will be reorganised and their staff redeployed. For the economy as a whole, delays in project implementation will be greatly reduced. Increased competition will lead to enhanced pressure on enterprises to reduce their costs and to improve quality.

The full outcome of the industrial policy changes will materialise after sometimes; but the early results are already visible. The response to the new industrial policy has been very encouraging. The number of investment approvals given in 1991 had risen to 5,538 from 3,335 in 1990.

Changes brought about in policies towards direct foreign investment have also evoked a strong positive response from foreign companies. In 1991, a total of 244 cases of foreign equity participation with a proposed equity investment of $ 504 strengthened our industrial capabilities and contributed to exports.

Reorganisation of Public Enterprises

The public enterprises were causing lot of drains on public exchequer. The recent researches showed that large monopolies, operating in the domestic market, without competition, tend to be inefficient and rigid in their response to changes in the pattern of demand or technological innovation. The success of countries like Korea and Japan in promoting productivity, exports and technological innovation is largely due to their success in realising the scale economies in production and learning, while at the same time enforcing competition and imposing a market discipline on large enterprises. Many of these successful enterprises are organised as coglomerates which operate in diverse fields and compete with each other. Korea has perhaps carried this strategy to the farthest by promoting and protecting the giant 'Chaebol'.

Programme of Economic Reforms

In 1991, India adopted economic reforms which had three basic components : (i) deregulation, (ii) competition, and (iii) strong Budget constraints. India thus joined 36 countries who have adopted Fund supported programme as dictated by the IMF and the World Bank. The broad thrust of the programme of measures is based on financial stability, outward booking policies and deregulation of markets.

There were three main points of the Indian reform programmes as introduced in 1991. First, fiscal deficits were to be reduced drastically. Second, exchange rates were to be depreciated and third, a number of structural reform policies were to be introduced, including tax reform, deregulation of prices, delicensing of investment, liberalisation of foreign trade and foreign investment regime, as well of public enterprises. On each of these points, it would have been possible even in 1991 to build-up a scenario of the expected outcomes of these policies in the next few years, in the light of the experience of the fund programmes in other countries.

For example, on the fiscal side it would have been quite predictable that the burden of deficit reduction would have fallen overwhelmingly on cuts in capital and development expenditures. Since the main thrust of the tax reforms was in the nature of reduction of rates and since simple Laffer-curve relationship between reduction of tax rates and increase in tax revenues does not exceed. It should have been expected that significant increase in revenues would take sometimes to materialize as the ability of the Government to increase tax coverage and tax compliance would be necessarily limited. The actual performance of the Government in raising tax revenue has been quite impressive and the credit must go to the efficiency of tax administration. Nevertheless, the tax-GDP ratio could not have been expected to rise very much and a reduction in the deficit GDP ratio would have to come mainly through reduction in expenditure. Since interest payments could not be waived, the only way primary deficits could be reduced was either by reducing administrative expenditures and subsidies or by cutting development and capital expenditure. Given the political realities in the country and also the fact that the beneficiaries from development and capital expenditures are too widely spread out to form a lobby, it was almost predictable that although the intentions of the policy-makers would be to reduce the revenue expenditure, in actual practice it would be the development expenditure that would suffer. Furthermore, as these development expenditures including public investment have a direct impact on the growth of output, which because of other structural policies was expected to be in any case sluggish, a deficit reduction through the reduction of development expenditures need not reduce the deficit GDP ratio as fast as the programme has intended.

Impact of Structural Reforms on the Indian Economy

The Finance Ministry outlined the following major reforms :

(i) to improve the health and viability of the Banking System;
(ii) reforms of Corporate Taxation and Customs and Excise;

(iii) management of public enterprises; and

(iv) emphasis on poverty alleviation and Social Service Programme with efficiency of delivery system and reducing leakages of funds.

The then Finance Minister, Dr. Manmohan Singh said : "Both in Banking and Stock Market Reforms, we have taken important initial steps. But we have an ambitious agenda ahead of us. Our objective is a vibrant, internationally competitive financial system which is adequately responsive to the needs of India's economic and social development."

Economic Reforms and the Poor: Reduction of all subsidies is not necessarily anti-poor, because quite often the subsidies support the relatively richer vested interests. But the reduction of development expenditures and public investment can have the visible impact of upsetting many of the projects, especially those in the social sectors such as health, nutrition, childcare and primary education, which affect adversely the common masses.

Financing of Social Sector

India needs to spend a lot more on providing quality education and health care services of her people. Yet today our ability to do so is greatly constrained by the large fiscal imbalances we are faced with. If we can contain subsidies and improve the rate of return on public sector investments, our manoeuvrability in this regard will considerably improve. The basic purpose of reducing fiscal imbalances is to enhance our ability to spend more in human resources development. The National Health Policy of 1983 committed the nation to spending 6 per cent of GDP on health care. The National Education Policy, 1986 talked of spending six per cent of GDP on education. The bulk of resources for education and health have to come from the public sector. This, in turn, requires a healthy fiscal system which can generate additional resources without frequent increases in tax rates. It also requires an economy which is growing fast enough. These are precisely the objectives of our stabilisation-*cum*-structural reform programme. If this programme is successfully implemented, we will have a lot more leeway in dealing more effectively with issues of social development.

Achievements

1. India had a record foreign exchange reserve to the tune of about $263 billion (approx) in Jannany 2010, *i.e.* Rs.12,15,318 crore. The foreign exchange reserve as on 2nd May, 2009 stood at $250 billion (approx), *i.e.* Rs. 11,80,538 crore whereas on 31st March, it stood at 241.4 billion U.S. dollar.
2. The gross fiscal deficit is just 6 per cent in 1994-95. In 1993-94 monetised deficit was 0.63 per cent and fiscal deficit was 7.29 per cent which is very good as compared to 1994-95, figures being 7.53 per cent and 2.62 per cent respectively.
3. The inflationary rates in April 1995 reached to one digit. From October 1990 the inflation rate rose rapidly to reach 13.7 per cent in February 1991. After a brief slowing down in April and May, this process of accelerating inflation continued until August 1991 peaking in the fourth week of that month at 16.7 per cent. Since then the rate of inflation declined by the end of January 1993 and had gone down to 11.8 per cent.

During the year 2008-09, India had exports worth Rs. 7,66,935 crore whereas imports were for Rs. 13,05,503 crore. But the import figure included import mainly of POL items. The import of petroleum product was only for Rs. 1,94,584 crore. Thus main emphasis in India should be to reduce dependence on import of petroleum products by creation of non-conventional sources of energy. Godavari basin is going to produce a large quantum of gas which will be used for fuel purpose and there will be little less dependence on petroleum. Even if we take that the production of gas will be worth Rs. 84,000 crore, yet it will not affect our import bill much. It is hoped that Indians will be able to tap new sources of petroleum either through joint ventures, exploration of new wells and basins in the country and non-conventional fuel like bio-fuel produced through jethropa.

The Changing Economic Milieu of India

To move onwards, to progress, to prosper is not only a fundamental right but also a cardinal duty of every person and every society. Inertia or stagnation harms both the society and the person but when the spirit of advancement galvanizes itself, when it leapfrogs with a desire to do something worthwhile and to show its calibre then the society also rises with enthusiasm and animation. When a work is executed with selfishness or compulsion, it is certain that people involved in doing the work will lack zeal for the task and then they find each step a heavy one to take and will vacillate in their purpose.

In the history of independent India, the first phase in the name of reformation began in 1969. This started an era of nationalization and nationalization of every industry was considered the cure of every malady. Such mania of nationalization was born on every occasion, without weighing the consequences, every industry, whether sick or healthy, essential or redundant, was annexed by the Government.

This epoch continued for more than a decade with the result that our public sectors became very heavy loss-incurring, inordinately expensive, perfectly lethargic and completely incompetent but this situation could not last for a very long time. Whispers were always there, smoldering for the alteration of this situation but due to political upheavals, change of Central Governments and new political priorities nobody could master courage to bring about the desired transformation. In the meantime, other Asian countries — China, Korea, Singapore, Malaysia, etc., who got their independence almost at the same time as India forged ahead on the path of progress; whereas India had started retracing its steps backwards. At the beginning of 1991 when oil crisis was born due to the Gulf War, India did not have enough foreign currency to import oil for more than two weeks. We had to pawn our gold to pay instalments of foreign debt. The speed of industrial development had become negative, unemployment and inflation were high and new social and economic difficulties started cropping up.

The way that we chose to rise from this situation lay in exactly the opposite direction from the one we had already been following. Earlier if Government controls were rigid then now they were suddenly finished. Earlier if personal asset building was looked down upon then now privatization became the magical formula of development and progress.

Just as at the core of the first phase of industrial and economic amendments, there was no ideological point of view, so this new age of liberalization also felt the lack of it. If first phase was entailed by political requirements, so this second period was the result of the discrepancies and complexity of the first one. Neither was absolute nationalization of our compulsion nor full-blown liberalization compatible with our social structure. Both the steps were forcibly taken and both hurt, in their own way, Indian entrepreneurship, talent and competition.

The most disturbing aspect of new economic policy was that it brought on the stage of world competition the Indian industry, which was nurtured for long under protection. It was the work of policy-makers to prepare Indian industry, physically and mentally, for this competition but instead of doing this, they opened all the ties and deprived Indian industry of Government protection. Thus the same Indian industry, which was cringing from public sector in the era of nationalization, now it is not only skeptical of the growing clout of multinational companies in this age of liberalization and globalization but has begun to weaken before them.

India does not lack resources to develop itself yet it is considered important to invite foreign capital. Foreign investors also have their eyes focused on the huge consumer market of India. If they will invest in India then their objective most certainly would be to multiply their profits rather than to fulfil the needs pertaining to development of India. However, this scenario could still be tolerated but when the Government, which is expected to protect the industry, employment and more than that the entrepreneurship and credibility of the country, begins to provide more facilities, privileges and priorities to foreign companies then a question mark could be seen hanging on the future of indigenous industry. A potent case in point here is one of the biggest public sectors of India, Bharat Heavy Electricals Ltd. (BHEL) which was being rendered moribund due to the policies of the Government.

Although Indian industrial world had not completely braced itself up to deal with this foreign economic invasion of Indian market yet it took courage and stood up in arms against the Multi-National Corporations (MNCs). Ramesh Chauhan's beverage manufacturing Parle company competed with Pepsi for a long time but eventually it had to make compromises. The same is the case with Kwality ice-cream. The name of this company still shows the popularity of its products but all the rights of this company now rest with 'Unilever'. Godrej had carved a niche for itself in the Indian market but due to this foreign invasion it had to make compromises with Proctor & Gamble. In the field of computers, Hindustan Computers Ltd. (HCL) had its own reputation but even this company buckled under stiff competition and had to strike a middle ground with Hewett Packard.

The form of these compromises could be anything. It can be called joint partnerships, giving the right to sell or it can be given the name of convenient business but the bitter truth is that those few Indian companies, which had earned a reputation with their dedication and diligence were forced to stoop before the MNCs. The better position of MNCs and the pressure of their money worked against the Indian companies and their products are either already being sold to the MNCs or are being compelled to do so. When big and renowned Indian companies are meeting with this fate, can the average industry stay on its own?

What is the reason that the long and firmly established Indian industry is falling like a pack of cards before the MNCs. As said before, the lack of foresight of Indian policies and decisions based only on the recent situations together form the main reason for this state of affairs. The policies, which once considered the Nationalization the miracle drug to boost Indian Industry, now deemed foreign investment as the panacea for curing the ills of Indian economy.

This, however, does not mean that all blame lies with the Government and its policies. The mistakes of the Indian private sector are no less. The private sector of India, for decades, had been charmed by the business of profit at the cost of improvement of the quality. Due to the protection given to the Indian industry by Licence policy, the consumer was obliged to purchase whatever was available to him at whichever price, but increasing competition gave him a chance to select the item of his choice from a wider range of products and this began defeating the Indian industry. In other countries, first a lot of money is spent on the R & D of new technology and then its benefits are reaped, while in India, we spend less than a half per cent of our National income on innovations and two-third of that money too, is spent by the Government. Our Labour Unions also do not flinch in demanding a pay-hike without increasing the productivity of the factory.

In those areas where advanced technology, skill and heavy investment are required and if there is a lack of these things in India, it is obligatory and important to invite foreign investment and technology. Inviting foreign capital to improve the basic infrastructure of Energy, Transport, Telecom etc. can also be spared a thought, but if we start depending on foreign investment for our common consumer goods then we will lose not only our industry but also our self-respect and eventually our self-reliance. We must remember that any person or society cannot survive if cut off from his roots for a long time.

On the other hand, globalization should be for the benefit of the Indian markets. If the benefits does not percolate down to the masses, there should be check on globalization specially flight of capital from India. It is promised time and again that the gains of globalization in the terms of taxes, wealth etc. will go down to the rural areas. The Indian regulatory system should be vigilant enough to ensure this and not to create further gulf between the rich and the poor, otherwise the Indian people will wash of the globalization from the Indian scene.

BIBLIOGRAPHY

Agrawal, P.K. (1996): *India's move towards sustainable development*, MD Publications Pvt. Ltd., New Delhi, pp. 125-131.

Current Statistics (2009): *EPW*, June 6-12, Vol. XLIV, No. 23, pp. 72-73.

Guha, Ashok (1990): *Economic Liberalisation, Industrial Structure and Growth in India*, Oxford University Press, Walton Street, Oxford Dx 2 6 DP, pp. 9-27.

Telegraph. English Daily Publisher for Kolkata, KG basin gas to fetch Rs. 84,000 crores, dated 22nd April, 2009.

World Bank, Idea (1997): Achievements and Challenges in Reading Poverty, A World Bank, Century Study, pp. III-XIV.

4

Globalization : Emerging Issues and Challenges for Developing Economies

Globalization is a process of economic, social and cultural integration of the World. Today globalization has become the fact of life in the World. It has emerged as the new international system that has replaced the cold war system.

Globalization is transforming trade, finance, employment, social system and ways of living, cultures and patterns of governances. Besides economic transformation, it involves changes in social and political structure and status of societies.

The real challenge has come from the forces of globalization. The multinational corporations, reinforced by the global economic and financial institutions like IMF, World Bank and World Trade Organization have made serious threats into the territorial, political, economic and cultural identities of the nations. This is more visible in the case of economically and politically weaker nations of Asia, Africa and South America. Globalization violates sovereignty and thus encroaches upon the basic rights of nations. The increase in consumerism and economic, social and technical inequality between the nations and also within the nations has compounded the problem of human rights.

It has widened the gap between the global centres of power and the rest of the World. The developing World has been the worst victims of globalization in a number of ways. They have been forced to take many anti-people measures like large scale reduction of workers, downsizing the Public Sector Undertakings (PSUs) or even selling them to private sectors. Rich nations have benefited from globalization. Their wealth and per capita income have increased. In other way now they are controlling the entire World resources directly or indirectly without any responsibility. Poor nations' resources are exploited by rich nations through technological know how.

Globalization is likely to adversely affect the growth of state sovereignty. At present, all Government organs, including the regional and local ones will have to listen to the dictates of the World Bank, IMF and WTO, as many of national level projects are totally controlled by the World Bank or MNCs in economic and financial matters.

There is an end of ideology and everywhere capitalism is dominating. Even the social movement is declining. The media, press, political parties, pressure groups and its agencies etc. directly or indirectly are part of globalization of western type, *i.e.* domination of one culture, one country, one military power and complete leadership of America and its allies, whereas, ancient Indian type of globalization is different. It is globalization of people instead of commerce only, it is based on the philosophy of '*Vasudhaiv Kutumbkam*', *i.e.* the whole world is one family. We know that the characteristic of family is co-operation while the modern globalization believes in the globalization of market. The characteristic of market is competition. So it is entirely opposite of the ancient Indian type of globalization.

The big players of the market, who are mostly governed by the profit motive, want to sell their products by adopting false and unreliable advertisement. In such a situation, the consumers are confused and it has become difficult for many of them to make right choice of products.

Today the middle class people in the developing countries are in the state of dilemmas and their loyalties are dependent upon technology, market and money which have been already globalized. The national affiliation of middle class, say in India, is replaced by a sort of corporate citizenship and to them now the Indian state is no good. Besides, one can trace the imprint of neocolonialism in the way the voice of the developing countries has been suppressed by WTO dominated by North. In the North dominated global trade, the developing states have hardly any voice. MNCs now play an overwhelming role in all spheres of life in the developing countries. Even in the remotest part of the World, one can notice the cultural invasion of neo-imperialism. The situation has come to such a level that South-South cooperation has practically been reduced to minimum. As an after effect of globalization, millions of people in the developing World not only face starvation, but also are victims of AIDS.

In the 1970s, modernization theory promised that developing

countries would "take off" if they opened their economies to free trade, that wealth will trickle down from rich to the poor. With the dawn of the new millennia, ironically enough, the rich are becoming richer and the poor are becoming poorer within the North and South. The glaring inequality between the rich and the poor countries is also reflected in the average annual income.

The Human Development Report, 1999 also had observed that 'gap' in income between the poorest and richest people and countries have continued to widen. Economic inequality and income disparities between and within the nations have contributed to rise in social strife and ethnic conflicts particularly in Asia and Africa. Today, poverty is the single most important factor responsible for human right violation. Now it is known fact that globalization instead of reducing poverty, it has increased the incidence of poverty, hunger and disease as the result of withdrawal of the state from the economic activities.

Rapid economic growth in several parts of the World has left a spillover effects on environment. Large scale of migration and urbanization has made worse that process. These days, deforestation, desertification, pollution, the depletion of ozone layer and resource depletion are not confined within the territorial borders of the nations. But obviously the developing countries are the worst hit. Urbanization also causes strains on the global ecosystem.

While globalization has worked well for a few, it has marginalized, dispossessed, deprived and destroyed many. While the rich and powerful have benefited, the poor and the weak have been excluded and defeated. Globalization, in this sense, it is not an inclusive phenomenon but an exclusive one. Globalization has resulted in loss of jobs, cultural identities, democratic rights, clean air and water. It has not only created social and cultural divide but also economic and technical divide between the nations and also within the nations.

Today, so there is no divergence of opinion that globalization has resulted in widespread of social and economic inequality. It has failed to solve the basic problems of poverty, hunger and unemployment, on the one hand and alienation, dehumanization, resource exhaustion, environmental pollution, ecological threats, violence and war on the other hand.

Economic Growth and National Integration

Generally, the terms economic growth and economic development are treated as synonyms, but they are not the same. Development should be related to developing countries while the term growth is related to developed countries. Economic growth means more output, *i.e.* quantitative expansion and economic development implies more output and changes in the technical and institutional arrangements, by which it is produced, *i.e.* expansion with qualitative changes. In spite of the difference we use both these terms as synonyms.

The main objective of the economic growth is to maximize the production of the consumer goods and services in the society, which increases the economic welfare of the people.

Economic growth is expressed and measured in terms of the growth of gross domestic product, *i.e.* G.D.P., per capita income and standard of living etc., but modern Economists have defined it in terms of economic welfare.

The United Nations Development Programme (UNDP) has presented the Human Development Index (HDI) for measuring the economic growth. Social indicators such as health, education, water supply, food, sanitation, dwellings etc. are the basic needs for the economic development.

Natural resources, human resources, suitable technology, high rate of capital formation, good organization and able entrepreneurship, financial stability and size of the market are the main factors of economic development but the non-economic factors play a major role in development. The main determinant of growth for a country is to have an efficient and stable Government because if the country is disturbed more frequently, economic progress cannot continue. A suitable socio-cultural atmosphere is necessary. In more developed countries, the progressive socio-cultural atmosphere is present and so they have higher growth rates, but underdeveloped countries have social institutions like caste-system, communal, apartheid system etc., which are not conducive to economic growth.

Right from the very beginning, the Indian state sought to promote total harmony among people professing different faiths, granting them freedom of worship and extending certain privileges to the members of the minority community. Despite the best efforts of farsighted leaders, communalism has blotted the fair name of the country known through ages as an oasis of tolerance and amity.

The people of India stand for a plural and multi-religious society. It would indeed be foolish to judge India's secular credentials and its innate capacity to tolerance by what happened recently in Gujarat, in

Ayodhya and Mumbai twelve years ago, and during the anti-Sikh riots in Delhi and elsewhere in 1984. These have been aberrations indulged in by petty minded people for immediate gains. The culpability of certain politicians and other elements including a section of media in stoking the embers of hatred cannot be ruled out.

A glimpse into the past would show that communal harmony and tenets of tolerance have been integral part of the great Indian tradition and unassailable thought flows through the Rigveda: "Truth is one the learned may describe it variously." The Atharvaveda says "The earth which accommodates people of different persuasions and languages as peaceful harmony benefit all of us." Echoes the Rigveda "All human beings are of the same race."

We see that when there is conflict among the people of the country on the basis of caste, religion, region or languages etc. the country becomes poorer and poorer because most of the time and money are involved in solving such problems.

In India, there are large regional variations in the level of development. It generates the pattern of low-sided growth and poses a danger to national integration.

It is true that all the regions cannot develop uniformly. For example, lack of suitable arable land or irrigation facilities may not congenial for spectacular agricultural development. There are large disparities in land, climate and mineral resources, irrigation facilities, level of education and other factor endowments among different states, which are bound to create economic disparities. All the regions, therefore, cannot be identically developed. In that case the Government should take every possible step for the most effective and fullest possible utilization of the available resources in that particular state so that overall growth of the economy can be enhanced. That much disparity is inevitable and hence not objectionable. What is objectionable is the neglect of a particular region at the cost of others. The experience is that in the planned era regional disparities have been growing. The landless agricultural workers, the marginal farmers and the poor tenants have been the worst sufferers. While the conditions of the rich, prosperous and big farmers have brightened up, the poor farmers have gone worse off.

The new agricultural technology was designed for those regions and areas that had adequate arrangements for assured supply of water for irrigation. All other areas were side-tracked. More favoured region were located in Punjab, Haryana, Western U.P., coastal areas of

Andhra Pradesh, Tamil Nadu, Maharashtra etc. The other regions and states were just like the spectators. As a result of this, regional imbalances emerged. The growing disparity in the rural areas accounts for much of the mounting discontentment and unrest among the rural population. Punjab and Haryana have developed due to green revolution while Maharashtra and Gujarat have developed due to industrialization.

The main cause of the Naxalites movement is improper distribution of land. With social justice, people will have more incentive for economic growth. The Government should help to obtain social justice through land reforms, reduction in inequalities of income, decentralization of economic power, workers' participation in management etc. This will bring institutional changes in the economy and promote economic growth of the country.

Atmosphere of peace and sound law and order situation in the country is very important for economic growth. It is primary duty of the Government to defend its frontiers from foreign aggression and also to protect life and public and private property within country. If there are regular disputes in the country on petty issues like language, caste and religion, state boundaries or terrorism etc., then no economic growth is possible. Our life on this planet is short. Let us learn to live together because we belong to the same family. Harmony is the key-note of the river of life.

5

Legal Reforms for Sustainable Development

Law & Justice

From the Middle Ages to the end of the eighteenth century, law was an aggregate of laws and a law was an authoritative rule of conduct for the individual man. This was the position adopted by Bentham[1], an English analytical jurist. The Soviet polity gave up the Marxian "withering away of law". Kelsen has argued the identity of state and law, saying that the State is a normative ordering co-extensive with the normative ordering of the system of law. Thus, the law slowly drifted from the authoritarian idea to the humanitarian idea. The idea of authority is taking on new form. Ever since the Reformation, the emphasis has been on change. The Reformation, the Puritan Revolutions, the English Revolution of 1688, the American Revolution, the French Revolution, the Russian Revolution – six major revolutions in four hundred years, or roughly one in every other generation – has made violent change seem the normal course of things and stability seem stagnations. Reformation emphasized on ethical element in law. Today philosophical methods in jurisprudence are employed to assess the ideal element in systems of developed law. Justice is the end of law. Justice is represented through body of philosophical, political and ethical ideas and may differ from time to time to some extent. This presupposes that the purpose of law is the administration of justice. According to Radbruch, justice is the ideal relation among men. The task of the judicial process is to strike a balance between individual freedom and community interest. The task of the judicial process is to maintain the legal orders as an order of justice by applying to decisions of controversies the authoritative norms of decision established or recognized by the State. Now-a-days courts talk about "the fundamental right to work for a living". Article 55a of the Charter of the United Nations, which binds the most of the nations of the World as member countries exhorts "higher standards of living, full employment and conditions of social progress and development." But it needs to be kept in balance with other ends. Roscoe Pound also talks of social engineering as end of law.[3]

He says that there is a real need for a field of human intercourse freed from legal restrain.

If one agrees to approach law from this standpoint, there is need for continuous legal reforms in the body of law in which the body politic and courts will have their roles to play. Of course, law made by courts through precedent cases is recognized as good law as law made by the legislation. Law should also recognize the rights, interests and values of the last man in the society. Misfortunes of majority should be borne by the few who are haves. The message of M.K.Gandhi is inscribed in the ground floor of UN Headquarters at New York saying that all developmental activities should aim the last have-not in the society. The real development should be able to wipe out the tear from the eyes of the poorest man in the society. Thus, the State is to promote human welfare to provide fulfilment for all men of their expectations in life in society apart from providing security or maintenance of law and order.

Legal Reforms

The term legal reform involves a critical study of the existing system of laws and administration of Justice highlighting the areas of deficiencies and improving upon them by plugging the loopholes.

Legal reform thus encompasses amendments in procedural and substantive law but would also include reformative and corrective measures aimed at strengthening and systematizing the law to make justice delivery system manifold and more effective.

Justice Hidayatullah said, "A Republic must possess the means for alternating and improving the fabric of the Government so as to promote the happiness and safety of the people.[4]"

The courts as part of modern republic or state also function as safety valve so that people's unrest doesn't cause violence or bloodshed in the society. Courts in their long journey, world over, have evolved their own means to invoke their inherent powers to discharge their onerous responsibility to the society as a safety valve or last refuge to the helpless or hapless individual or group of individuals harassed and driven away by the might of the state. Thus, a legal system to be organic and responsible to people's need, should have well greased mechanism for reforms. Most of the polities do have the system of reform. But reform has its own philosophy or ideology. Reforms mostly lead the society but sometimes reforms are caused due to strong ventilation of public grievances or due to vigilant public consciousness or opinion. However, both should have harmonious balancing. Reforms are for the people, of the people and by the people. Therefore, the reforms should be implemented considering the absorbing power of the populace. Reforms which go over the heads of people are only converted into intellectual luxury best preserved in books kept in libraries. If reforms have come after reading pulse of public, they will be welcome and will have willing support of people and will bear fruits in the practical way. Thus, legal reforms should keep pace with the sustainable development in the society. There is no contradiction between the two. In fact, sustenance of law presupposes welfare of the people. In few cases like in dictatorship or in guided democracy, the legal reforms will depend on the will of the dictator or the coterie controlling the republic. In such cases, legal reforms will have to wait for the good sense or wisdom of the persons at the helm of affairs but there will be no stopping of it. None can stop the smooth flow of truth or compassion. Truth will ultimately win over all obstacles whichever come in its way. If generally accepted principles of life cannot find suitable expression or patronage, violent revolution will be natural consequence. Of course, there might be some delay because a consensus has to be arrived at. However, law without enforcement has no meaning. In ultimate analysis, law making and law reforms are two sides of the same coin.

Legal reform is an strong instrument to ensure equitable justice. For example, land reform laws, if implemented faithfully, can ensure equitable distribution of landed property, which is the main stake of the rural people without any financial involvement. Similarly, taxation laws ensure taking away of extra income from the wealthy to the less fortunate fellow-citizens. Environment protection laws came into force in plenty world over after Stockholm Conference in 1972. The Conference asked to take proper steps for the preservation and regeneration of natural resources of the earth. It was pointed out at the Stockholm Conference that poverty itself is the greatest environment pollutant. Sustainable development is development that not only generates economic growth but distributes its benefits equitably, that regenerates the environment rather than destroying it, that empowers people rather than materializing them.

Let us examine how the legal system has helped in achieving the object of everlasting development and prosperity to the people. The Constitution is the ground norm of each country. We shall see how legal system has been addressed the problem or achieved sustainable development in various countries.

In China, Article 19 of the Constitution of People's Republic of China declares that all citizens have the right to work. Thus, State accepts its responsibility for providing work and for improving conditions of work. Article 100 says that citizens of China must uphold discipline at work. As per Constitution of China, the aim of the judiciary in China is to give to the citizens of China, speedy and cheaper justice. But for all practical purposes, the judiciary is subservient to the Legislature, the Executive and the Procurators and cannot dare give decisions against them. All natural resources belong to the State by law.

The Constitution of Japan guarantees the fundamental human rights to the people which shall be eternal and inviolate rights. The enjoyment of the freedoms and rights guaranteed to the people by the Constitution shall be maintained by the eternal vigilance of the people and the people shall refrain from any abuse of these freedoms and rights, and shall always be responsible for utilizing them for public welfare. All people shall be respected as individuals and their right to live, liberty and the pursuit of happiness shall, within the limits of the public welfare, be the supreme consideration in legislation and the Governmental affairs. Social equality guarantees the abolition of special privileges. A large measure of local autonomy is given so that by way of decentralization, democratic institutions may take roots in the country.

The Constitution of erstwhile Burma which came into force in 1948 was abrogated in 1962 and at present the country is being governed by decrees issued by the Government. In such a situation, it depends on the good sense of the dictator. However, the State is to direct its policy towards exploitation of all natural resources by itself or local bodies or by people's cooperative organizations.

In Sri Lanka, Parliament has power to make laws for the peace, order and good governancy of erstwhile Ceylon. There is no prohibition or restriction on free exercise of any religion. Local Government up to the village level are autonomous under the Central Government. However, sovereignty of the people is inalienable.

The Constitution of Pakistan (1956) guaranteed certain fundamental rights to the people of Pakistan like right to equality, protection against retrospective offences etc. The Constitution of 1962 and of 1973 provided principles of law-making. The principles of law making included that no law was to be repugnant to Islam. The policy of law-making in Pakistan, apart confirming to Islamic principles, is based on good principles of polity like the fundament rights in India. It even provides right to work and employment. But the validity of an action or of a law shall not be called in question on the ground that it is not in accordance with the Principles of Policy. Secondly, the Judges of the Supreme Court or High Court can be removed if the Supreme Judicial Council recommends to the President his physical or mental incapacity or he is found guilty of misconduct. Any legal reform, in Pakistan thus, depends on Principles of Policy and will of the President who is most of the time an Army General by force.

The Constitution of Bangladesh is based on high ideals of nationalism, socialism, democracy and secularism. It has fundamental rights and basic principles like the Constitution of India.

Australia has a Parliamentary form of Government. The High Court of Australia is the highest Federal Court of the country. Judges are appointed for life to the various courts and they hold office during good behaviour. The Constitution of the State can be amended by the Legislature of the State by ordinary law.

Perusing various systems of fundamental laws in some of the South Asian countries, it is obvious that the constitutions are quite flexible to undertake legal reforms. They contain basic good principles of polity. The problem is with regard to enactments by the Legislatures to implement sustainable development. Even if the Legislatures pass enactments, the performance of various Acts, Statutes or laws can be called in question. The problem is more of implementation than law-making. However, many countries have still to pass laws preserving their air, water, soil qualities so that the defaulters can be duly taken to task by the right of the state. There is no effective impact of public opinion or public censure on forest produce, smugglers, water and air polluters. It is where the judiciary has also intervened. Judiciary in most of the countries feels its sacrosanct obligation to protect and preserve environment. "Law has to be interpreted according to the current societal standards. The law when enacted, in spite of the best efforts and capacity of the legislators, cannot visualize all situations in future to which that law requires application. New situations develop and the law has got to be interpreted for the purpose of application, for the purpose of finding a solution to the new problems. This is how the law advances. That is the area or field of judicial creativity to fill the gap between the existing law and the law as it ought to be."[5]

Judicial review was first conceived in the United States by Chief Justice Marshall of the US Supreme Court in 1803 in *Marbury* Vs *Madison*, but really the seed of that was sown two centuries earlier in 1608 by Lord Coke in *Bentham's case*. At that time, some treated it as sacrilege. Nazis under Adolph Hitler reduced the Weimer Constitution to a nullity by kidnapping the requisite number of the opponents in the Legislature and obtained needed majority and gave all powers to Hitler including the power of amendment of the Constitution. Thus, Rule of Law became a joke and all development was set at rest what to say of sustainable development. This type of legal reform is not warranted by good conscious and is not recognized by the World community at any cost at any time whatsoever. Karl Marx, a visionary of twentieth century warned that human activities, which disregard valuable rules of nature only invite disaster. This message electrified Soviet leaders. They took lead to promote eco-movement.[6]

He advised that we should leave a better environment for the coming generation. This principle was reiterated in details in the conference of 1992 in Brazil on Sustainable Development. Leader of the United States, President Theodore Roosevelt had shown similar concern. He issued a note of caution, "The natural resources are in danger of exhaustion if the old methods of exploiting them further continued. The time has come to inquire seriously as to what will happen when our natural resources are exhausted."

Both the USSR and the USA though the environment they had. was not seriously in peril, showed a serious concern. In their perception, natural sources are the national assets. Costa Rica had unique and very significant approach. The country, in 1948, abolished its national army and diverted entire Budget for the cause of environment. Preservation of environment is required but social justice without economic growth is impossible. Prosperity is to be created. Clear environment has no meaning to the tribals of forest rest areas or aboriginals of America and Australia without means of sustenance or maintenance of living for them. Poverty is the biggest pollutant so also unlimited greed of the advanced societies. 'Economic growth without Social Justice is inhuman', says Toynbee.[7]

> "Now, more than anything else, economic development provides the basic analytical framework of the UN Effective assistance to the development of the least developed countries (LDCs) to attain the 'welfare world' upgrading the living standard of the lowest stratum of international society has been gradually becoming a priority programme of the UN."

About 80 per cent of the UN Staff are engaged today in the solution of problems of economic and social development.[8] Poverty has caused unprecedented AIDS in African and Asian countries. Where fight on poverty or AIDS is to be launched on economic or political fronts, simultaneously legal measures will be required to contain both. Wealth generated around the World through information technology would have been cornered by the hackers or manipulators, had the Acts to control these malpractices not been in place by various countries in due time. So also, the polluting industrial units would have spoiled water and air for the human beings had they were not compelled by legal measures to take effective measures. In the present day regulated world, law reforms or measures are the safest and easiest way to ensure long time benefits from development by harmonious balancing.

The price regulated for certain things like specific fish, air-conditioner, cars etc. would prevent people from going for uncontrolled purchase of that thing which causes environmental hazard. Similarly, expenditure taxation will lower material living. Carbon Taxation will result into less carbon emissions. Strict Environment Impact Assessment (EIA) procedures will change investment decisions and major development projects will make provision for rehabilitation of evictees or oustees. No purpose will be served from a big development project if number of banefited evicted persons is the same as number of beneficiaries benefited after a long gestation period of the project. Reforms through regulations or law can tackle these problems to a large extent. "In SAARC countries—Bangladesh, Bhutan, India, Maldives, Pakistan and Sri Lanka—the direct migration of rural folk to the large cities and metropolises has agglomerated a large number of them who could hardly be absorbed gainfully in the urban economy."[9] They leave paradise like natural habitat in Nepal, Bhutan and Maldives in search for jobs and livelihood. Thus, economic policies should have proper linkages with ecology concerns. Uncontrolled growth would surely result in large scale depletion of natural resources and disturbance of ecological balance. That suggests need for setting up of the International Environment Authority, which will regulate only three major ecological concerns in the beginning, *i.e.* acid rains, green-house effect and depletion of the ozone layer. This will also coordinate the activities of Inter-Governmental Organization (IGO), NGOs and States.

I agree with Shri P. Puneet[10] that International Law is not enforceable law yet I don't agree with him that an International Environment Court should be established as it will amount to compromise with the sovereignty of the member states who are only competent to reform and enforce laws within their own territories and none else can take up this task.

Indian Scenario

India has been one with nature from the time immemorial. Therefore, the Indians have always been happy people. The Fundamental Rights and the Directive Principles of the State Policy together forms the bulwark for sustainable development for any country. Article 39 (b) and (c) envisages for socialistic pattern of the society. Article 51A (g) states:

> "Every citizen has a duty to protect and improve the natural environment including forests, rivers and wildlife and to have compassion for living creatures."

To achieve above mandate, the plethora of legal reforms through legislation or executive law making have been carried out. In fact, the environment statute book has become too heavy to be grasped by the citizens and even by the persons who are behind their implementation unlike in other South Asian countries. There are at least two hundred different statutes in India on environmental pollution control. Out of these, at least fifty statutes are on water pollution control alone which is the main pollution problem of India. The Supreme Court has taken the lead in this field and has pre-empted or precipitated many legal reforms which may be termed even as judicial activism. In *Rural Litigation & Entitlement Kendra* Vs. *State of U.P.*[11], the Apex Court ordered for stoppage of unauthorized and illegal mining in Mussoorie-Dehradun belt. Many may criticize it as over enthusiasm of the Hon'ble Court but one could find perceptible improvement in the environment of the road from Dehradun to Mussoorie and one cannot but bless the Apex Court from his heart.

In *Banwasi Seva Ashram* Vs. *State of U.P.*[12], the Supreme Court held that legal aid should be provided to the claimants of the compensation. Without adequate facilities by NTPC the persons cannot be evicted from their land or habitat.

In *M.C. Mehta* Vs. *Union of India*[13], the Supreme Court while protecting the pristine glory of the river Ganga, held: "Just like an industry which cannot pay minimum wages to its workers cannot be allowed to exist, a tannery which cannot set in up a primary treatment plant cannot be permitted to continue to be existence for the adverse effect on the public at large." The Court further observed, "We are conscious that closure of tanneries may bring unemployment, loss of revenue, but life, health and ecology have greater importance to the people."

However, the law should provide solution to the retrenched workers. Similarly, the Supreme Court was able to reduce air pollution in Delhi by compelling plying of CNG buses and autoes in Delhi and sale of lead free petrol but the administration has been allowing few small-scale industries from the backdoor. This system is still awaiting successful implementations in other metro-cities. The Supreme Court has emphatically stopped degradation of forests from the nature-bounty North-Eastern region in spite of clamour by the timber merchants, industrialists, smugglers and by some of their men or agents. Through Public Interest Litigation, the courts are now dictating many matters of public concern. Some advise the courts to observe self-restraint. The courts have to amend their own home first which is cracking under the load of unprecedented arrears of cases numbering in crores. A reform legislation carried out or propounded or pronounced cannot be implemented if it is not practical or the society is not ready to accept it. Justice A.S. Anand, former Chief Justice of India has, therefore, advised to remember following observations of Lord Justice Lawton in *Laker Airways*[14] [1977(2)

WLR 234 at 267] "Policy is determined by Ministers within the legal framework set out by Parliament. Judges have nothing to do with either policy-making or the carrying out of policy. I regard myself of a referee, I can blow my judicial whistle when the ball goes out of play, but when the game restarts, I must neither take part in it nor tell the players how to play."

In the epoch making judgement in *Kesavananda Bharti*[15] case, Justice Dwivedi opined from the dissenting side as below:

> "The Constitution does not recognize the supremacy of the court over Parliament.
> The Constitution is not intended to be the arena of legal quibbling for men with long/heavy purse. It is made for common people."

As the situation stands today, the subordinate courts are not in a position to decide cases including all important environment or pollution cases before 3-4 years time, if not more. Setting up of environment courts or earmarking one in each district of the country can be one of the solutions. The Law Commission of India has also recommended for creation of Environment Courts or Green Courts from where appeal will lie only to the High Courts or the Supreme Court. The power to impose fine, disconnection of power, closure of industrial unit is not enough through regulations as regulations are not so powerful. Secondly, laws should be so framed that one is always loser after being convicted for violation of any provision of the environment related Act, rule, regulation etc. like the Environment Protection Act, 1986. The E.P. Act should have maxim 'polluter to pay' expressly. All environmental laws should have basic approach that polluters should pay specially in view of precarious financial position of most of the State Governments in the country. Hon'ble Supreme Court has already been pleased to lay down polluter to pay principle in connection with W.P. (C) No. 914 of 1996 (*Vellore Citizens' Welfare Forum* Vs. *Union of India & Ors.*) and also reiterate in connection with Matter No. 3727 (C) of 1985 (*M.C. Mehta* Vs. *Union of India & Ors.*). Relevant portion of the order reads as follows :

> "Some of the salient principles of 'Sustainable Development' as called out from Brundtland Report and other international

> documents, are inter-Generational Equity. Use and Conservation of Natural Resources, Environmental Protection, the Precautionary Principle, Polluter Pays Principle, Obligation to assist and cooperate, Eradication of Poverty and Financial Asistance to the developing countries. We are, however, of the view that 'The Precautionary Principle' and 'The Polluter Pays' Principle are essential features of 'Sustainable Development'........"

"The Polluter Pays" principle has been held to be a sound principle by this Court in Indian Council for *Enviro Legal Action* Vs. *Union of India* JT 1996 (2) 196. The Court observed, "We are of the opinion that any principle evolved in this behalf should be simple, practical and suited to the conditions obtaining in this country." The Court ruled that "Once the activity carried one is hazardous or inherently dangerous, the person carrying on such activity is liable to make good the loss caused to any other person by his activity irrespective of the fact whether he took reasonable care while carrying on his activity. The rule is premised upon the very nature of the activity carried on." Consequently the polluting industries are "absolutely liable to compensate for the harm caused by them to villagers in the affected area, to the soil and to the underground water and hence, they are bound to take all necessary measures to remove sludge and other pollutants lying in the affected areas. "The Polluters Pays" principle as interrupted by this Court means that the absolute liability for harm to the environment extends not only to compensate the victims of population but also the cost of restoring the environmental degradation. Remediation of the damaged environment is part of the process of "Sustainable Development" and as such polluter is liable to pay the cost to the individual sufferers as well as the cost of reversing the damaged ecology.

"The precautionary principle and the polluter pays principle have been accepted as part of the law of the land." Similarly, there should be restriction on the legislatures so that they do not spend for populist purposes binding future generations for their political gains at present. The accountability of bureaucrats or executives should be fixed. They should be made responsible for results in respect of preservation or improvement in environment and not only maintaining formalities or procedures. The Legislatures with the able assistance from the executive and support from the judiciary just after independence have shown through implementation of integrated reform of land laws that law reforms is not only the instrument for sustainable development but can also prove as an end itself. Supporting land reform measures, Justice Hidayatullah stated:

> "Grotius had treated the right as an acquired right and ownership as either serving individual interest or for public good. According to him, the acquired right had to give away to eminent domain but there must be public interest and if possible, compensation. In the Social Contract theory also, the contract included protection of property with recognition of the power of the ruler to act in the public interest and emergency.[16]"

Through introduction of Panchayati Raj System in the country after its inclusion in Part-XI-A of the Constitution in 1992, the rural people have woken up to their problems and will prove in due course of time that they can look after themselves. Even 33 per cent rural women representatives have proved their skill and mantle to run local self-government. I am sure that through many more such legal reforms towards gender sensitization, uniform civil code, administrative law etc., the sustainable development will no longer be a dream but will be a reality in India.

NOTES

1. Bentham, (1979): *Principles of Morals and Legislation* (ed), 324.
2. Kelsen, (1934): Reine Rechtslehre, chap.8.
3. Pound Roscoe, (1958): *The Ideal Element in Law, Tagore Law Lectures*, 1948, published by University of Calcutta, p. 370.
4. *I.C. Goloknath & Ors.* Vs. *State of Punjab & Another*, AIR 1967 S.C., 1643.
5. Justice J.S. Verma (1997): *Equality, Liberty and Prosperity under the Indian Constitution*, Eastern Law House, Calcutta.
6. Desai, Ashok A. (2000): Constitutional Accountability towards Environment, in *Journal of the Indian Law Institute*, 42 JILI, Bhagwan Das Road, New Delhi-110001.
7. Toynbee, Dr. Arnold (1976): *We the People* by N.A. Palkhiwala, Strand Book Stall, Mumbai, p. 73.
8. Chakraborty, Bimal (1997): *The United Nations and the Third World – Shifting Paradigm*, p. 266, Tata McGraw-Hill Publishing Company Ltd., New Delhi.

9. Ganguly, J.B. (1976): 'A Strategy for Sustainable Development of India', p. 180, in *Sustainable Development Strategy* (ed.) By S.P. Shukla & Nandeswar Sharma, published by Mittal Publications, New Delhi.
10. Puneet, Pabby (2005): 'Global Environment Change & International Law' in *Environment Development*, (ed.) Jabir Singh, I.K. International Pvt. Ltd., New Delhi.
11. AIR 1988 SC 2187.
12. AIR 1987 SC 374.
13. AIR 1988 SC 1037.
14. Anand, A.S.: 'Judicial Review – Judicial Activism – Need for Caution', p. 158, published in *Journal of the Indian Law Institute*, 42, JILI (2000), Bhagwan Das Road, New Delhi-110001.
15. *Kesavananda Bharati* Vs. *State of Kerala & Another*, 1973-74, SC Cases, para 1947.
16. AIR 1967, SC, pp. 104-105.

BIBLIOGRAPHY

Agrawal, P.K. (1996): *India's Move Towards Sustainable Development*, MD Publications (P.) Ltd., 11, Daryaganj, New Delhi-110 002.

Agrawal, P.K. (1993): *Land Reforms in India, Constitution and Legal Approach*, MD Publications, New Delhi.

Chakraborty, Bimal (1997): *The United Nations and the Third World, Shifting Paradigms*, Tata McGraw-Hill Publishing Co. Ltd., New Delhi.

Kashyap, Subhas C. (2003): *The Citizen and Judicial Reforms* (*Under Indian Polity*), (ed.) Universal Law Publishing Co. (P.) Ltd., G.T. Karnal Road, Delhi-11003 (under the auspicious of Centre for Policy Research, New Delhi).

Mahajan, V.D. (1986): *Select Modern Governments*, S. Chand & Co. Ltd., Ramnagar, New Delhi-110 055.

Handbook & Journal

National Law School of India (2000): *Handbook on Human Rights for Judicial Officers*, University of Bangalore.

Verma, S.K. (Ed.) (2000): *Journal of the Indian Law Institute*, Vol. 42, No. 2-4, April-December 2000, Bhagwan Das Road, New Delhi-110001.

6

Impact of Global Financial Crisis on Indian Economy

Introduction

Financial crisis seems to have been occurring with greater frequency in the era of globalization. The four major crises were in Latin America in the early 1980s and Mexico, Asia and Russia in the 1990s. The fifth one is the recent global financial crisis. Ten years ago, financial crisis of the East Asia was due to a real estate bubble burst in Thailand, triggering the flight of international speculative capital. Today, it is fallout of the real estate crisis in the USA, which threatens the financial markets. The current crisis differs from the Asian crisis also because the search of its origins is not leading to some form of crony capitalism in emerging economies but to the centres of the World financial system with all modern financial innovations. Some analysts say that this is first crisis in the modern financial markets where new products created by debt and asset securitization are traded globally.

The current crunch shares certain qualities with prior financial crisis, such as the pattern of rising asset prices, expanding lending activity, increasing speculation to excess and then declining prices, insolvency and finally, panic in which the 'investors' try to save their gains as rapidly as possible.

The global financial crisis of 2008-09 emerged in September 2008 with the failure, merger of several large United States based financial firms and spread with the insolvency of additional companies, Governments in Europe, recession and declining stock market prices around the globe. But the financial crisis really started to show its effects in the middle of 2007. Around the World, stock markets have fallen, large financial institutions have collapsed or been bought out and Governments in even the wealthiest nations have had to come up with rescue packages to bail out their financial systems.

The crisis has become one of the most radical reshaping of the global banking sector, as Governments and the private sector battle to share up the financial system following the disappearance of Lehman Brothers and Merrill as independent entities. Actually, the collapse of Lehman Brothers was a symbol of the global financial crisis. The real economy in many countries is already feeling the effects. Many industrialized nations are sliding into recession if they are not already there. The crisis became so severe that after the failure and buyouts of major institutions, the Bush administration offered a $700 billion bailout plan for the US financial system.

This bailout package was controversial because it was unpopular with public, seen as a bailout for the culprits while ordinary persons would be left to pay for their folly. However, as Nobel Prize winner for Economics Joseph Stiglitz argued that the plan "remains a very bad bill". He said that this bill is again based on "trickle down economics." You throw enough money at Wall Street and some of it will trickle down to the rest of the economy. It does not do anything about the basic source of the problem.

Some of bailouts have also been accompanied with charges of hypocrisy due to the appearance of socializing the costs while privatizing the profits. The bailouts appear to help the financial institutions that got into trouble while smaller business and poorer people rarely have such options for bailout and rescue when they find themselves in crisis.

On the one hand, many people are concerned that those responsible for the financial problems are the ones being bailed out, while on the other hand, a global financial meltdown will affect the livelihoods of almost everyone in an increasingly interconnected World.

A collapse of US subprime mortgage market and the reversal of the housing boom in other industrialized economies have had a ripple effect around the World. Furthermore, other weaknesses in the global financial system have surfaced. Some financial products and instruments have become so complex and twisted, that as things start to unravel, trust in the whole system started to fall.

The global economy is on the dangerous situation of recession. The downturn after four years of relatively fast growth is due to a number of factors : the global fallout from the financial crisis in the United States, the bursting of the housing bubbles in the US and in other large economies, increasingly restrictive monetary policies in a number of countries and stock market volatility.

There is the argument that when the larger banks show signs of crisis, it is not just the wealthy that will suffer, but potentially everyone. With an increasingly interconnected World, things like a credit crunch can ripple through the entire economy.

Market liberalization and privatization in the commodity sector have not resulted in greater stability of international commodity prices. There is widespread dissatisfaction with the outcomes of unregulated financial and commodity markets, which fail to transmit reliable price signals for commodity producers.

After the Second World War, forty-four nations designed the Bretton Woods system of international finance to help reconstruct and stabilize a post-war global economy. In the 1970s, the purpose of these International Financial Institutions (IFIs) shifted towards a neo-liberal economic agenda, championed by Washington (also known as the Washington Consensus). It was at this time that policies such as structural adjustment started to be pushed too much of the developing World, following a "one size fit for all" prescription of how economies should be structured, which had disastrous consequences for much of the World's population.

Although such institutions have rarely been held accountable for such policies and their effects for many years, people have been calling for their reform, or even for their abolition. The old institutions have recognized the need for reform, but they have been moving in a glacial speed. They did nothing to prevent the current crisis.

It took the World fifteen years and a world war to come together to address the weakness in the global financial system that contributed to the Great Depression. It is to be hoped that it will not take us that long this time; given the level of global interdependence the costs would simply be too high.

For the developing countries like India, the rise in food prices as well as the knock on effects from the financial instability and uncertainty in the industrialized nations are having compounding effect. High fuel costs, soaring commodity prices together with fears of global recession are worrying many analysts.

Impact of the Crisis on the Indian Economy

The Indian economy has shown negative impact of the present global financial crisis. Though the public sector in India, including nationalized banks, could somehow insulate the injurious effects of globalization as we are also part of the globalization strategy of neo-liberalization, there is a limit of our ability to resist global recession, which may change into a great depression.

The impact of the crisis will be significantly different for the Indian economy as opposed to the Western developed nations.

Table 6.1 : Trends in GDP at Factor Cost in Rs. Crore

Year	*GDP (at 1999-2000 Price)*	*Growth in %*
1999-2000	17,92,292	–
2000-01	18,64,773	4.4
2001-02	19,72,606	5.8
2002-03	20,48,287	3.8
2003-04	22,22,758	8.5
2004-05	23,88,384	7.5
2005-06	26,12,847	9.4
2006-07	28,71,120	9.9
2007-08	31,29,717	9.0
2008-09	33,51,653	7.1

Source: Central Statistical Organization, Government of India.

After a long spell of growth, the Indian economy is experiencing a down turn. Table 6.1 shows that in 2006-07, the GDP growth rate was 9.9 per cent which became 9.0 per cent in 2007-08 and due to the impact of recent global financial crisis and global recession, the growth rate of Indian economy is now declining. In 2008-09, it was reduced to 7.1 per cent. The International Monetary Fund (IMF) has also projected the growth prospects for the Indian economy to 5.1 per cent in next year. This declining trend has affected the service sector adversely. Service export growth is also likely to slow as the recession deepens and financial services firms, traditionally large users of outsourcing services are restructured. The financial crisis in the advanced economies and likely to slow down in developing economies could have adverse impact on the IT sector. About 15 to 18 per cent of the business coming to the Indian outsources includes projects from banking, insurance and the financial services sector which is now uncertain.

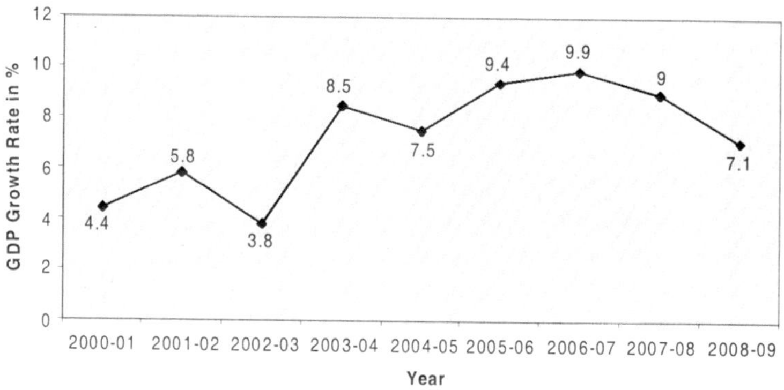

Figure 6.1 : Trends in GDP (Growth Rate in %)

A financial crisis can cause workers' earnings to fall as jobs are lost in formal sector, demand for services provided by the informal sector declines and working hours and real wages are cut. When formal sector workers who have lost their jobs enter the informal sector, they put additional pressure on informal labour markets.

Industrial growth is faltering. India's industrial sector has suffered from the depressed demand conditions in its export markets, as well as from suppressed domestic demand due to the slow generation of employment. As per the Index of Industrial Production (IIP) data released by CSO, the overall growth in April-November 2008 is estimated at 3.9 per cent compared to a growth of 9.2 per cent in April-November 2007.

The recent crash in the Sensex is not simply an indicator of the impact of international contagion. There have been warning signals and signs of fragility in Indian finance for sometime now, and there are likely to be compounded by trends in real economy.

The most immediate effect of that crisis on India has been an outflow of foreign institutional investment from the equity market. Foreign Institutional Investments (FIIs), which need to retrench assets in order to cover losses in their home countries and are seeking havens of safety in an uncertain environment, have become major sellers in Indian markets. As FIIs pull-out their money from the stock market, the large corporate will no doubt be affected, the worst affected are likely to be the exports and small and marginal enterprises that contribute significantly to employment generation.

In 2007-08, net Foreign Institutional Investments (FIIs) inflows into India amounted to $16040 million. But in April-November 2008 it was negative to $8857 million (Table 6.2).

Table 6.2: Net Investment of FIIs at Monthly Exchange Rate (in US $ Million)

Year	*Amount*
1999-2000	2339
2000-01	2160
2001-02	1846
2002-03	562
2003-04	9949
2004-05	10272
2005-06	9332
2006-07	6707
2007-08	16040
2008-09*	–8857

*April-November, 2008-09.

Source: Security and Exchange Board of India (SEBI).

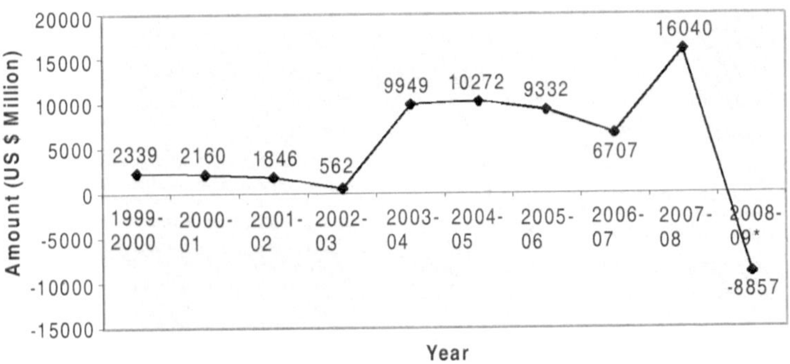

Figure 6.2 : Net Investment of FIIs (At Monthly Exchange Rate)

Due to this, there is a collapse in stock prices. As a result, the Sensex fell from its closing peak of 20873 on January 2008 to nearly 8000 in October-November 2008 (Fig. 6.3).

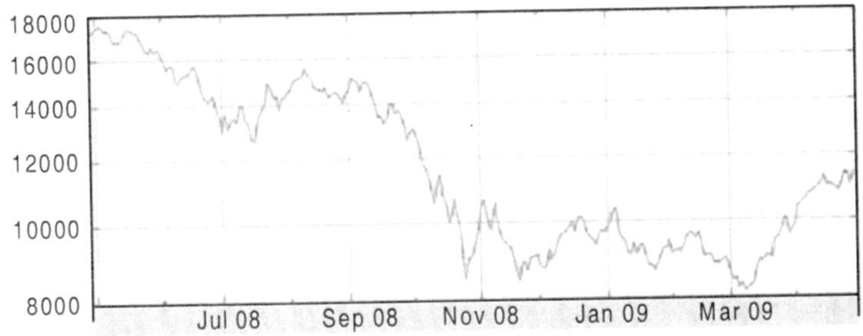

Figure 6.3 : Daily movements of BSE Sensex in 2008-09

Source: Bombay Stock Exchange, India.

In addition this withdrawal by the FIIs and corporate converting the funds raised locally into foreign currency to meet their external obligations led to sharp depreciation of the rupee. Between April 2008 and November 2008, the RBI reference rate for the rupee fell by nearly 25 per cent, rupees per unit dollar gonc up from Rs. 40.02 in April 2008 to Rs. 49.00 in November 2008 (Table 6.3).

The currency depreciation may also affect consumer prices and the higher cost of imported food hurt poor individuals and households that spend much of their income on food.

Table 6.3 : Foreign Exchange Rate

Month	Rupees per unit of Dollar	Appreciation/ Depreciation
March 2008	40.36	–
April 2008	40.02	+0.85
May 2008	42.13	–4.2
June 2008	42.82	–5.74
July 2008	42.84	–5.79
Aug 2008	42.91	–5.95
Sep 2008	45.56	–11.42
Oct 2008	48.66	–17.05
Nov 2008	49.00	–17.64
Dec 2008	48.63	–17.01

Source: Monthly Economic Report, Ministry of Finance, Government of India.

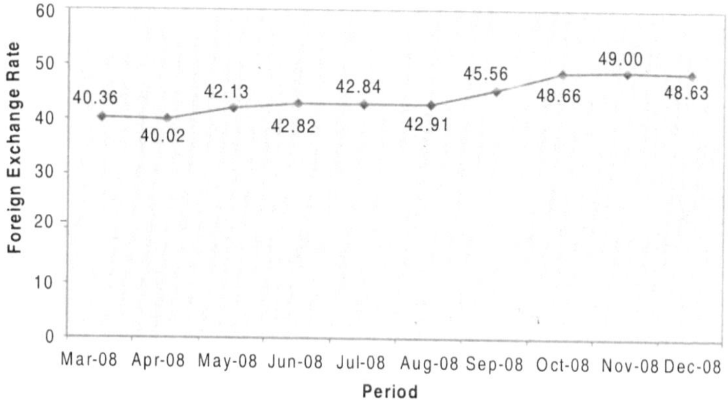

Figure 6.4 : Foreign Exchange Rate (Rupees Per Unit of Dollar)

Figure 6.5 shows that the rate of inflation has gone down to 8.98 per cent in the last week of November 2008 from the peak of 12.9 per cent in first week of August and further sharply declined during recent few weeks. The recent trends suggest a faster rate than expected reduction in inflation. The decline in inflation should support consumption demand and reduce input costs for corporate. From the external sector perspective, it is projected that imports will shrink more than exports keeping the current account deficit modest. But the current account deficit is widening.

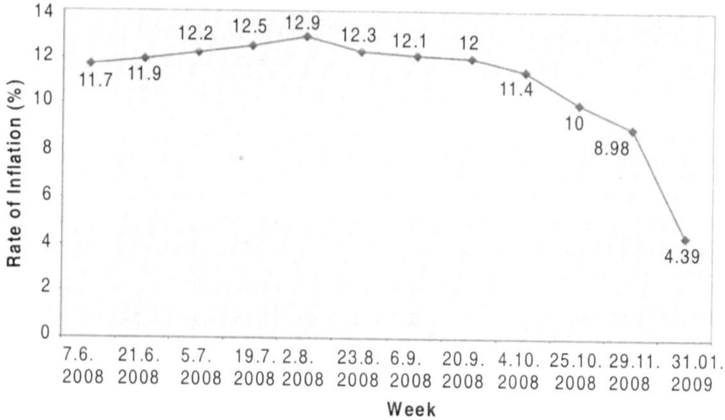

Figure 6.5 : Rate of Inflation (WPI) in India

Source: Office of the Economic Advisor, Government of India.

The foreign exchange market came under pressure because of reversal of capital flows as part of the global decelerating process. Foreign exchange reserves are depleting. It was $309161 million in 2007-08 and now in 2008-09 it is $247621 million, which shows the direct impact of the financial crisis on India's foreign exchange reserves (Table 6.4).

Table 6.4 : Foreign Exchange Reserves (in US $ Million)

	2001-02	2002-03*	2003-04	2004-05	2005-06	2006-07	2007-08	2008-09*
Foreign Exchange Reserves	54106	75428	112959	141514	151622	199179	309161	247621

*As on 23rd January, 2009.
Source: RBI Weekly Statistical Supplement.

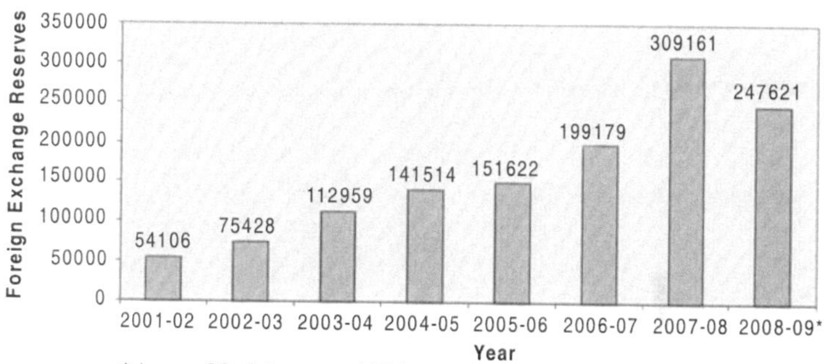

*As on 23rd January 2009.

Figure 6.6 : Foreign Exchange Reserves (in US $ Million)

The shrinking of aggregate in the World market as a consequence of the crisis has hurt the exporting manufacturing industries in the country. Table 6.5 shows that in 2007-08, India's exports and imports were $155512 million and $235911 million respectively and balance of payment was $ –80398 million. And now exports and imports are $131990 million and $225809 million respectively. The balance of payment is $ –93819 million. This shows that India's exports are adversely affected by the slow down in global markets. This is already evident in certain industries like the garment industries where there have been significant job losses with the onset of the crisis. This along with a squeeze in the high income service sectors like financial services, hospitality, tourism etc. will lead to a reduction in consumption spending and overall demand with the domestic economy. A direct consequence of this is a simultaneous loss of informal employment and lower generation of new non-farm employment in the economy. The depreciation of rupee could not positively affect the exports bill of India.

Table 6.5 : India's Foreign Trade (in US $ Million)

	2002-03	2003-04	2004-05	2005-06	2006-07	2007-08	2008-09*
Exports	52719	63843	83536	103091	126414	155512	131990
Imports	61412	78150	111517	149166	185735	235911	225809
Balance of Trade	–8693	–14307	–27981	–46075	–59321	–80398	–93819

*April-December 2008-09.

Source: DGCI&S, Kolkata.

The other direct impact of the global financial crisis will occur in the area of credit availability to the small-scale agriculture and other rural livelihoods. The impact of the crisis on the rural sector will originate from the slow down experienced by secondary and tertiary sectors. The fact that the present crisis will adversely affect the manufacturing and service sectors imply that occupational diversification will be more difficult to achieve. The financial crisis, therefore, threatens to intensify the income deflation that is already a feature of the rural economy and simultaneously aggregates the alarming levels of hunger and malnutrition that currently exist in India. Lately, the crises seems to be over in India yet the basic causes still exist.

How the Government and RBI Responded

The Government of India and RBI responded to the challenge strongly through its fiscal and monetary policies.

The Government has introduced three fiscal stimulus packages. These are expanded safety-net programmes for the rural poor, the farm loan waiver package and payout following the Sixth Pay Commission report for stimulating demand in the economy.

On the other hand, in aftermath of the turmoil caused by bankruptcy, the Reserve Bank has announced a series of measures to facilitate orderly operation of financial markets and to ensure financial stability, which predominantly includes extension of additional liquidity support to banks. The RBI has been effectively able to manage domestic liquidity and monetary conditions consistent with its monetary policy.

This has been enabled by the appropriate use of a range of instruments available for liquidity management with the Reserve Bank such as the Cash Reserve Ratio (CRR), which is reduced to 5 per cent in July 2009 from 9 per cent in August 2008 and Statutory Liquidity Ratio (SLR) stipulation and Open Market Operations (OMO) including the Market Stabilization Scheme (MSS) and Liquidity Adjustment Facility (LAF). Reduction in the repo rate (the rate at which RBI lends to the banks) from 9 per cent as in October 2008 to 4.75 per cent in July 2009 and the reverse repo-rate (RBI's borrowing rate) reduced to 3.25 per cent in July 2009 from 6 per cent as in October 2008 in order to improve the flow of credit to productive sectors at viable costs so as to sustain the growth. Furthermore, money market liquidity is also impacted by our operation in the foreign exchange market, which in turn, reflects the evolving capital flows. The existing set of monetary instruments has thus, provided adequate flexibility to manage the evolving situation.

So the financial sector has emerged without much damage. Thanks in part to our strong regulatory framework and in part on account of most of the nationalized banking sector.

Conclusion and Suggestions

No doubt, India has been hit by the global crisis. It is clearly due to India's rapid and growing integration into the global economy. The strategy to counter these effects of the global crisis on the Indian economy and to prevent the latter from any further collapse would require an effective departure from the dominant economic philosophy of the neo-liberalism. The first such departure should be a return to Food-First doctrine, not only to ensure food security of the large population but also due to the fact that food production will be more profitable given the current signs of a shrinking market for export oriented commercial crops. The other important initiative that needs to be adopted is the building of institutions based on the principle of cooperation that will provide an alternative framework of livelihood generation in the rural economy as opposed to the dominant logic of markets under capitalism. Institutions like cooperative markets and credit cooperatives can go a long way in addressing the lack of economically viable producer prices and loaning credit availability for economic activities in the primary sector. Such an alternative policy to tackle the consequences of the financial crisis will require effective Keynesian policies in the form of increased public expenditure at the rural and urban infrastructure. We see that Government's engagement generally arrives very later to solve the financial crisis by which time many financial firms are near insolvency. This generates larger cost for the economy and exchequer. Our key goal today should be to avoid these costs through rapid action.

The need of today is not just the pumping of liquidity into the Indian economy but also in addition the injection of demand. This can occur only through direct fiscal action by Government. In India, larger government expenditure has to be oriented towards agricultural, rural development, health, human resources and infrastructure to make inclusive and balanced growth.

The new growth will have to come not from some new speculative bubble but from enlarged Government expenditure that directly improves the livelihoods of the people and that is geared towards improving the production of foodgrains through a changing of peasant agriculture and not through corporate farming since that would reduce purchasing power in the hands of the peasantry and perpetuate its distress. In short, the new paradigm must entail infrastructure and foodgrain-led growth strategy on the basis of cultivator agriculture sustained through larger Government spending towards the agriculture and rural sector, which can simultaneously remove both recession and food crisis in India.

REFERENCES

Akyuz, Yilmaz (2008), "The Global Financial Crisis and Developing Countries," *Resurgence*, December, Penang, Third World Network.

Athukorala, P. and Sen, K. (2002), *Saving, Investment and Growth in India*, Oxford University Press, New Delhi.

Central Statistical Organization, Government of India.

Chandrasekhar, C.P. and Ghosh, Jayati (2004), *The Market that Failed : Neoliberal Economic Reforms in India*, Left World Books, New Delhi.

Chandrasekhar, C.P. and Pal, Parthapratim (2006), "Financial Liberalization in India: An Assessment of its Nature and Outcomes", *Economic and Political Weekly*, March 18.

International Monetary Fund (2008a), *Global Financial Stability Report*,

Washington D.C. October.

International Monetary Fund (2008b), *World Economic Outlook*, Washington D.C., October.

Monthly Economic Report, Ministry of Finance, Government of India.

Monthly Statistics of Foreign Trade of India, Vol. I & II (Various issues), Director General of Commercial Intelligence & Statistics (DGCI & S), Ministry of Commerce, Government of India, Kolkata.

Panagariya, Arvind (2008), *India: The Emerging Giant*, Oxford University Press, New York.

Patnaik, Prabhat (2007), "Financial Crisis, Reserve Accumulation and Capital Flows", *Economic and Political Weekly*, December 15.

Patnaik, Prabhat (2008), *The Value of Money*, Tulika Books, New Delhi.

Reserve Bank of India (2008), *Annual Policy Statement for the Year 2008-09*, April. Reserve Bank of India, Mumbai.

Reserve Bank of India, *Weekly Statistical Supplement*, Reserve Bank of India, Mumbai.

Security and Exchange Board of India (SEBI), Mumbai.

7

Private Sector Development Transforming Life

Private Sector development is a reality now. Public Sector enterprises were earlier thought to be baseline for sustenance of the economy and the general mass of a country. After the collapse of the Russian economy and adoption of capital path by China, the public sector is no longer considered as the only way out. On the other hand, the World Bank, IMF and other international finance agencies had to come for rescue of many developing and undeveloped countries. In fact due to mismanagement of their economy there was a windfall. The popularly elected Governments had no resources to provide goods and services to back-up their electoral promises. As a result, they had to look elsewhere. On the other hand, the capital in developed countries were awaiting flight for different destinations where they could survive. "It was becoming extremely expensive to keep workers employed and pay for their upkeep even in times of recession or to keep up inventories of stocks, raw materials and so on. "The decade of the 1990s had more than its share of tumultuous political and economic events, beginning with the collapse of the Soviet Communist bloc and the huge problems of transition of the earstwhile Communist countries to the capitalist system and their integration with the World Trading system, the decade-long stagnation of the Japanese economy, the East Asian miracle and the East Asian crisis and the debt crises of Mexico, Brazil and Argentina. On the other hand, there was industrial slow down in 22 industrial countries. The growth rate declined from 3.8 per cent in 2001 to 1.1 per cent in the end of 2001. World exports and imports also declined by 3.56 per cent and 3.46 per cent respectively. GDP growth in the US slowed down in 2001 to 1.3 per cent. The real GDP stagnated in Germany and Japan and unemployment increased. Even some workers in Germany took wage-cuts to prevent their capital from moving out to Asia. It was becoming extremely expensive to abide by labour and environmental standards in the developed countries as part of international trade agreements. Therefore, capital decided to remain in places where it could 'farm out work' and simply keep control in its own hands. This explains why in 1980s and 1990s manufacturing units got relocated to the Third World countries in a big way and it was only much later that services and call centres started moving out. However, a closer look reveals that this process '....of capitalising' is rather contingent and not a universal law."

Private Sector has been long working in the field of power, health, urban infrastructure, telecommunication, education and research and even in defence in developed countries. But all results are not OK. There are lot of complaints of high power tariffs, high cost of goods and services, unhealthy competition and even blackmailing by private companies of the consumers. Consumers have been fighting a grim battle against gaps in their promises and actual services provided. But all consumers cannot fight and are not willing to fight out. They are busy in their highly competitive workplace. On the other hand, they fear another kind of exploitation by professionals like lawyers and middle men. It is so far satisfying that the law is widely respected and clearly implemented in the countries like USA, Australia, UK and other Euro-countries but the peace loving Japanese would not even think to knock at the door of watchdogs of law. It is heartening to know as on date that every conscious Japanese is good in his work place also, yet one has to provide for contingencies when things turn ugly due to excessive greed of people and lust for goods and services. The developed countries so far have maintained their services by ensuring high costs earmarked for these services. These high costs can be sustained only in the commanding heights of the economy. One really feels shaky to ring up from Tokyo or to Tokyo as the cost of call is too high. It is a fact that commanding heights of these economies cannot be maintained unless these economies look for markets elsewhere in less developed countries or developing countries and exploit their purchasing power. The purchasing power of the developed countries has reached a saturation point as technology cannot replace man. The man as machine has to function. All transactions need not be automatised. Thus, search or hunt for new markets will be the focus of the twenty-first century economic vehicle.

After public resentment to locate industries causing pollution in developed countries, focus has been shifted to the locations in developing countries and in the less developed countries. Due to information technology revolution, the message of NGOs of developed countries has reached the other countries. Secondly, industries which have been functioning so far have not perceptibly improved the standard of life of the local people. The person who is employed in an industrial unit generally gets separated from his original family and gets settled in new environment fending himself and his nucleus family. Thus, the industrialization has not in fact won the hearts of general population. They feel that their fate is first linked with their traditional agriculture or trade or business. They cannot calculate the indirect impact of industrialization on their life style nor there is adequate brainwashing or publicity. Thirdly, the Governments in the developing and less developed countries are only instruments for their welfare activities. The administration and management in Governments have become so slack, corrupt, inefficient and indifferent that they are not able to become efficient conduits to carry gains of improvement in health of economy or better revenue return or collection to masses. Therefore, masses remain where they were. An apathy is, therefore, developed among masses towards industrialization. They feel more secured in their small trade, business or agricultural farm. More the revenue is generated, more is non-developmental expenditure on bureaucracy in the Third World countries. The bureaucracy has become so well entrenched and unionized that it extracts most of gains of liberalization or globalization in its own fold as a pressure group without increasing its efficiency. Therefore, one is compelled to doubt alongwith Karl Marx whether industrialization generates employment.

Globalization as a process is not bad. Globalization cannot be avoided. The moot question is how to ensure growth in the era of globalization or liberalization with justice. Foreign Direct Investments (FDIs) is a major factor in the Chinese economy in 1990s, after the beginning of the liberalization of the Chinese economic regime. Moreover, China had also taken care of her agriculture and demonstrated quite extraordinary rates of growth of her agriculture. It is, therefore, prerequisite that FDIs should be invited but at the same times domestic competition should be built-up so that FDIs don't run away with huge profits or remittances to their countries of origin. "In India, most people have either very low income or very low skills to be affected by globalization either way except software field. Low incomes imply, that imports don't swamp domestic markets thus protecting workers somewhat. And imports are and should be allowed in manufacturing sector and not in consumer or agro-products. Low skills imply that Indian workers cannot, under present arrangements, take advantage of greater integration of the Indian economy with rest of the World. Clearly, integration brings threat and opportunities. In the context, low buying power wards off the threat to the Indian economy.

Production from multinationals will allow low skill opportunities utilized. However, Indian could marginalize herself by not engaging with the rest of the World. Let us take up some case studies in India regarding impact of private sector on the masses.

(A) Cola War in Plachimada, Kerala

Coca-Cola, one of the biggest multinationals in the World, had set up its plant in Plachimada, a small village in the district of Palakkad, Kerala in March 2000 in the wake of liberalization. The multinational more accused by the local people of creating severe water shortage in the area of polluting its ground-water and soil, and also of distributing toxic waste as fertilizer to farmers in the area.

Responding to the mounting pressure from the local community – particularly tribals the *dalits* (downtrodden) – the local Panchayat refused to renew Coca-Cola's licence for further production. The State Government also imposed ban. The Single Bench of Kerala High Court upheld the decision of the State Government. The Division Bench of the Kerala High Court has ruled in favour of the Coca-Cola company.

The High Court passed its judgement on the premise that it is unfair to refuse to quench thirst of Coca-Cola Company whereas it was given permission at the first place after considering all pros and cons.

The local community formed the People's Struggle Committee. The large number of NGOs and other organizations supported the movement both within the Court and out on the streets.

The matter is pending in the Apex Court. It says that after the 73rd Constitutional Amendment and inclusion of Parts IX in the Constitution, the Court cannot take away their power in controlling their local resources.

This agitation will spread in other parts of India where multinationals have located their units. This would clear the way for a new understanding of the problem and might give the campaign a new sense of direction which was largely led by the downtrodden themselves without any outside support. Perhaps if the tribals were to speak for themselves – whether in court or in the public sphere – the impact would be far more decisive. Legal niceties and judicial procedures would then be rendered redundant in the face of harsh realities in these times of neo-liberal reforms and cultural globalization.

(B) Public-Private Partnership in Urban Infrastructure Sector

It has been well established that private sector intervention has been most successful in development of urban infrastructure specially in real estate development. The impact of liberalization of economy is first seen in real estate field. This is a field where the private sector feels most secured as it can earn its return by controlling land and then selling of buildings or facilities to public after negotiation with them. Once the private developer has been given ownership and conversion of land, he becomes the owner of the land which none can challenge. This has to a large extent solved the complex housing problems in metro cities and also in middle class towns. Its visible impact is seen after seeing the improvement in quality of life as people stay in a better habitation or environment and fulfil their cherished desire to live in better accommodations. At the same time, the private sector has not been so successful in development of other urban infrastructure because the returns on public infrastructures are not so assured and depend on so many macro level conditions and political environment. The urban regulations are controlled either by the local body, state body politic or by the Central or Federal government. Therefore, it is not a smooth ride. Even some successful stories are there in this field. One such is of the Bangalore Agenda Task Force (BATF) which wanted to develop city infrastructure of Bangalore which is a metro city of India and is known as 'Silicon Valley of India' all over the World. The aim was to make Bangalore a "World class city". It was through entry of private sector into higher levels of policy-making. It was a relationship between business elites and government. It was to be based on financial efficiency and reducing subsidies while supporting infrastructure development for the controlled growth of urban areas. It was to satisfy demands for higher living by the emerging middle class in India which is comparable with middle class anywhere in the World now.

Public Private Partnership (PPP) has been adopted as a panacea to overcome failures of development and governance in developing countries by World Bank, Asian Development Bank, United National Development Programme and USAID and the Government of India adopted PPPs as a critical element of urban reform policy.

The Task Force constituted, had the erstwhile Chief Minister of Karnataka S.M. Krishna, the then Commissioner of the Bangalore Development Authority and the CEO of Infosys Corporation. The BATF engaged in a range of activities towards urban esthetics, planning and service provisions. BATF identified the following main seven 'stakeholders' in the city, Government to partner with :

1. Bangalore Development Authority (BDA),
2. Bangalore Water Supply and Sanitation Board (BWSSB),
3. Bangalore Electricity Supply Company (BESCOM),
4. Bharat Sanchar Nigam Ltd. (BSNL) – telecommunication provider,
5. Bangalore Metropolitan Transport Corporation (BMTC),
6. Bangalore City Police, and
7. Bangalore Mahanagar Palika (BMP).

A contribution of rupees 13 crore by private trust of Mrs. Sudha Murthi and Nilekani's Trust helped to their consultants from India and abroad including one from France.

The project helped in modifying accounting system within BMP. In other organizations, it provided the opportunity to bring financial accountability and performance measurement into the public space and to act as catalysts in the larger process of bringing the Government and the public closer together. The policies, plans and works were discussed among administration, corporators and grassroots communities. This is also the spirit of Part XI A of the Constitution added by the 74th Amendment Act of the Constitution. The public or Government contribution comes mainly in the form of land acquisition and subsidy. Total land about 26,392 acres was proposed to be acquired for Bangalore, Mysore Highway Project, International Airport, IT Corridor (Rs.18,000 crores). However, the public rejected the innovation as S.M. Krishna Government was voted out of power from Bangalore itself because the poor strata of Bangalore city felt them further marginalised and deprived. As a result, the BATF got a jolt. New Government, which came in power on a promise to do away with this innovation, later on paid only lip sympathy. But the BATF model was adopted by Government of India through JN NURM (Jawaharlal Nehru National Urban Renewal Mission) and is being adopted and replicated for selected middle and metro cities of the country.

(C) Public-Private Partnership in Health Sector in India

Health is another sector where the private sector has come up successfully so far in developing countries. This was necessitated because Governments failed at various tiers to provide satisfactory level of services to the people. There is notable absence of medico and para-medico personnel in remote areas. In 1991, the number of doctors per one lakh population in India was 47.19 only. Of all hospitals and beds in the country, only 32 per cent and 20 per cent respectively were in rural areas. Therefore, the rural populace is left with no other option except to turn to private practitioners, quite often to trained Unani doctors or in non-allopathic systems like Ayurvedic and Homeopathic system or go to local quake or 'ojha' or flock to already crowded block, subdivision or distinct level hospitals.

Public Private Partnerships (PPP) in health sector as practised in some of the states in India are given below:

(i) Offering subsidized land;
(ii) Certain exemption on excise duty on import of machinery as well as other fiscal incentives to private sector to set up super specialized hospitals;
(iii) Handing over management of primary health centres to NGOs; and
(iv) Involving industries in managing health centres.

Let us study a PPP model in Madhya Pradesh which is working under the banner 'Rogi Kalyan Samiti' (Patients' Welfare Committee) (RKS).

The RKS seeks partnership of all concerned actors at the local level to raise funds for the upkeep of the public hospitals and in turn ensures their participation in the management of health services. The RKS seeks direct involvement of the users (people) and service providers (doctors, para-medicos) in running the public hospitals.

The RKS was reportedly formed in 'more than half of the nearly 1200 public hospitals in Madhya Pradesh' and an estimated Rs. 370 to

400 million rupees was raised across undivided Madhya Pradesh. The patients are charged admission fee for the health care services they avail except from people Below Poverty Level (BPL). RKS triggered further innovative steps such as allowing commercial use of hospital land.

The local rich people or their organizations contributed liberally for this public cause and local monitoring committee is formed either with local MLA as Chairman or local Panchayat President or municipal chairperson. It is apparent that moribund Government hospitals can be made effective with little innovations. The successful functioning of RKS will help change the mindset that the Government is all doer and service provider and also those services could be availed without paying or sharing the cost.

Thus, health services can improve by providing space for the community centred actions, encourage experiments and encourage people in meaningful participation.

(D) Electricity Reforms

Electricity reforms started from USA. Except in few states like California, the private sector has engaged successfully in production and distribution of power in USA. The high cost of power is the only point that bothers to the policy planners and the consumers alike.

However, in developing countries like India, there is no other alternative. In view of the high cost of generation and inefficient functioning of States' Electricity Boards in the field of distribution, there is no other alternative except to resort to the participation of private sector in this vital sector of public importance. This is what was resolved in the "Washington Consensus".

However, privatization of the power sector with an independent regulatory like in the telecommunication sector, can solve the problem.

The Electricity Act, 2001, frees generation, captive generation from licences except for hydro units. It allows open access to transmissions, licensing for transmission and distribution will continue. There will be State level Regulatory Commissions (RCs) and retail tariff will be determined by them.

This has created competition which is the mother of all efficiency which was earlier absent in Government run power sector. In fact, that sector had become the green pasture of the power engineers to make easy money and was under the clutches of the politicians to fulfil their interests either of self or of their party. Thus, inefficiency and corruption were writ large in the sector and a gloom had descended over the power sector.

In developed countries, the fully restructured model involves vertical disintegration and deregulation. To safeguard interests of consumers in India, it is essential in absence of well organized consumer-movement in the country, to make regulatory commissions with multi-members and arm them with such overbearing powers and authorities that powerful lobby or cartel of private power sector players cannot play with the interests of innocent consumers.

It is also to be kept in mind that this is towards ultimately privatization of power sector. Therefore, progressive improvement of public services should be in-built in the process of liberalization of the sector. Competition for competition sake has no meaning unless it delivers good to the people at the grassroots level. Thus, there should be focus on improving governance free of corruption and incentive mechanism in the sector. Equal emphasis should also be placed on energy efficiency and alternative energy technologies. In 1990s, the economy was consuming 30 per cent less energy than the level in 1970s. The reforms should address the issues such as more access to power, less dependence or imports and other social and environment issues.

India should look into long-term sustainability rather than short-term political priorities. If it is not done, the country will be left behind rest of the World including China and Brazil.

(E) Setting up of Manufacturing Unit : A Case of Small Car Project of M/s. Tata Motors in Singur, West Bengal

About 1000 acres of land was acquired by the West Bengal Industrial Development Corporation for the Small Car Project of M/s. Tata Motors Pvt. Ltd. in 2006. Tata is the most reputed industrial house in India. It was a privilege for West Bengal Government to host the plant of Tata Small Car Project to manufacture Rs. 1 lakh dream car for the Indians. This was expected that with the arrival of Tata, other big industrial houses and companies will follow suit in the industrial starved State of West Bengal, which had otherwise earned an ill-reputation of labour problems and lack of work culture.

Govternment of West Bengal decided to acquire land in record time for the Tatas on their conditions and provided them many concessions in land acquisition costs, sales-tax etc. The land was acquired without studying the pulse of local people and without the knowledge that it was the multiple cropping land and was one of the best for agricultural purposes in the State.

While the land was acquired, the opposition in West Bengal opposed it tooth and nail saying that the farmers would not give away their lands as these lands were only sources of their livelihood. As a result, farmers of 234 acres of land out of 1000 acres (approx.) did not accept the compensation amount. This became the cause of ever simmering discontentment. They formed an organization under the banner 'Save farm land'. The supporters often disturbed the ongoing construction work and demolished the boundary wall many times. But the strong political will of the Government of West Bengal duly supported by M/s. Tata Motors, did not surrender to the demands of the local people to return back their fertile lands and instead use the land in front of the proposed unit which was less fertile.

The battle for the political supremacy had the following unwanted consequences :

(a) One emerging young woman supporter of 'Save farm land' was raped and murdered and this offence was committed at the instance of the ruling CPI (M) elite.
(b) There were cracks among the partners of the Left Front Government, which had been well cemented so far and has been ruling in West Bengal for more tham thirty years.
(c) There was a protracted court battle in the High Court of Calcutta through the Public Interest Litigation.

The problem of Singur attracted the World attention to the complex debate of land acquisition *vis-à-vis* development. If the local populace doesn't desire development, then why the development work is thrust on them. This also brought out the lax attitude of the West Bengal bureaucracy and the political elite who had been vouching as the only spokesman of the poor in the State. The development work was undertaken even without providing for a well laid down rehabilitation and resettlement policy even in the Communist regime of the Left Front Government. Ultimately, Tatas withdrew from the project and are in the process of vacating the site. Government of West Bengal has been left with no alternative.

The agitation against the setting up of the car manufacturing plant has cautioned the administrators and the policy makers that they should not take for granted the support of the poor. They are not only the deciders of their destiny. Their fate or progress has to be well chalked out in consultation with the local people at the grassroots level. Even elected panchayats or local bodies cannot take for granted their support. The support is only issue based and election mandate does not give sole right to the elected representatives to speak or do everything on behalf of their electorates. The common people have no faith that they will be employed. They cannot support jobless growth. The land evictee should be trained in skills whereas they can get a job of their liking. The industrial house or management should provide jobs for the land-losers and should make provisions for due compensation considering development in the area as a result of settling up of the industrial unit. The manufacturing unit should plan for people's needs and aspirations, to run the plant successfully in future. An unit cannot run smoothly without the support of the local people. These are the important lessons from the Singur land debacle.

(F) Agitation Against Proposed Industrial Hub: A Case of Nandigram, West Bengal, India

Nandigram is a very interesting case study which brings out the fact that apprehensions can also cause large scale public agitations and violence.

The chronology of the events run as follows :

Pursuing the declared policy of promotion of industries, Government of West Bengal had declared unofficially that there will be a chemical hub and other housing projects near Haldia Industrial Centre under Haldia Development Authority. For that purpose, land will be acquired to the tune of more than 5000 acres. Rumours were having their sweep. In the meantime, the local political MP (Member of Parliament) being the most influential person, got published a pamphlet through a semi-Government organization Haldia Development Authority that they would acquire land in 10-12 villages near Nandigram. This created panic among the villagers and they grouped under a non-political organization known as 'Save Land Eviction Committee' irrespective of party affiliation.

They even restricted entry of Government officials to those 10-12 villages where the land was supposed to be acquired. They cut roads and put up barricades. Once, they were cutting Government *pucca* road, the police intervened. They stoned police force and in the melee, one police sub-inspector was killed and was thrown in the nearby Haldi river. The police force was sore about the incident and wanted to take revenge. Arrests on large scale were not possible because of resistance by about ten thousand people of the villages. Their anger was severe on CPI (M) supporters. As a result, the CPI (M) supporters were compelled to leave their houses. They took shelter in their nearby party bastion village Khajuri. All ten villages covering a population of about thirty thousand were united for one cause that the takeover of their land should be opposed at all costs even at the cost of their lives.

When a large police contingent under a senior officer of DIG rank went to arrest the culprits in earlier cases filed, the villagers started pelting stones and put up obstacles on the roads leading to their villages. As a result, police resorted to bursting of tear gas shells and then firing as the crowd became uncontrollable and became very violent using bombs, grenades and arms against the police force. The ruling party supporters joined hads with police force and used brickbats, arms and bombs against members of 'Save Farmland Committee' which was locally known as 'Resistance Group against land eviction' (Bhumi Uchhed Pratirodh Committee). In the melee, 16 people died including women who were opposing the land acquisition. Many policemen also got injured and were admitted in various hospitals.

This became a major issue at the State and the National level. The 'Left Front Government' in the State went on the back foot. The Chief Minister declared that the so-called notice/pamphlet issued by the HDA for the proposed industrialization stands cancelled and withdrawn. Henceforth, land acquisition will be done only on approval of State Government through one agency, *i.e.* the Land & Land Reforms Department of the State Government and Collector, Purba Medinipur. The mass killing was criticized even by all National leaders. Government of India declared that no fertile land will be acquired for Special Economic Zones and this will be left on sponsorers to mostly purchase land for SEZs.

The conclusions of the Nandigram fiasco were the following:

(i) Cases of land acquisitions should be scrutinized closely by the higher echelon of administration and only minimum land required for the project should be allowed to be acquired.

(ii) As far as possible, fertile agriculture land should not be acquired. If the industry or unit cannot be placed elsewhere, then alternative land should be arranged or offered for the cultivators to cultivate in nearby area even though the compensation paid to the land evictees.

(iii) The people of the area should be heard properly before land acquisition plans are finalized.

(iv) Land acquisition evictees should be given proper compensation even considering future increased value of the land due to industrialization or development.

(v) Land-losers should be given jobs in the new unit as far as possible through a well-represented body set up for the purpose.

(vi) No land acquisition proposal should be cleared by the authorities unless it has proper rehabilitation and resettlement plan.

Conclusions

Thus, we should have a society with large sector of agricultural production that work in tandem with a certain kind of industry, rather than are dominated by an abstract thing called industry – irrespective of whether it produces poison, nuclear hazards or automobiles. It is a misconception that in developing countries, private sector can replace the public sector in one go. It is to be softly done because the populace is not ready to absorb the reform, culture and are doubtful of the private sector due to their past history and practices. The South-East Asian countries had gone fast on private sector involvement and their economies were at the brink of collapse in nineties. On the other hand, China and India have been involving the private sector cautiously. Any reform without people's involvement will be lopsided. It is bound to fail ultimately. For example, uncontrolled land acquisition for industries is being opposed violently in India by masses specially for setting up of Special Economic Zones, are large manufacturing units on fertile lands. The multinationals should not also be allowed to exploit raw materials of developing countries like India without ensuring improvement of quality of life of people. This type of development has become development terrorism. One cannot afford to have oasis in the poor neighbrhood. Equitable distribution of wealth still holds good and justice should reach to the last man in the society and should be a part of his holistic development.

In developed countries, it is an integral part of their liberal or market economy whereas in developing countries, it has to be side by side with public sector enterprises. Even in developed countries like USA, Japan and Germany, power sector and health sector privatization has not provided requisite relief and satisfaction to the consumers. On the other hand, the policy-makers cannot afford to go for regulated wholesale and unplanned entry of private sector in all fields of economy as people in developing countries are habituated to Government run services along with regulations. A popular Government can only bring private sectors fully by taking risk of its political fortune. Private sector involvement will depend on country's absorptive capacity in terms of human capital, depth of domestic financial market, quality of governance and macro-economic policies.

The best way out is public-private partnership and not the unfettered growth of private sector. It is a fact that the private sector will replace the PPP model growth soon because of their efficiency and speed. They will create jobs and prosperity for the people and people will welcome them. In due course of time, they will also become exploitative and will purchase public sector which will be engrossed with corruption and inefficiency. An inefficient organization has no place in the society. Only more efficient organizations will rule in the economic sector. But the popularly elected Governments will have to put such regulatory mechanism in place which will function as watchdog on the private sector. The private sector is not an end in itself. They will also exploit the consumers if they are not controlled. Industries are for people and not people for industries. They should work for co-operation and participation of local people. The private sector is welcome provided it is transparent and efficient and delivers good to the people. Ultimately, welfare of people is the mission of all organizations together and they should work jointly and tirelessly to achieve that mission to avoid public resentment, violence and even blood revolutions.

8

Role of Service Sector in the Indian Economy

Following the conventional classification, an economy is divided into three sectors, that is, agricultural (or primary), manufacturing (or secondary) and service (or tertiary) sectors. The agricultural sector consists of farming, forestry, animal husbandry and fishery. The manufacturing sector is composed of construction and manufacturing. All other economic activities, which are not covered by the agricultural and manufacturing sectors are broadly defined as services and hence belong to the service sector. They include services provided for agricultural sector, activities associated with the supply of water, electricity and gas, transport and telecommunications, wholesale and retail trade, finance and insurance, business and personal services and community and social services. Services can be broadly distinguished between two types, that is, the old and new services. The old or traditional services include trading, domestic services, catering and hotel services. The new services are generally associated with communications, business and legal practices, software services, research and education. These are generally tradable internationally and hence are also called tradable services.

Since the early 2000, the World has witnessed the spectacular growth of the Indian economy. Associated with this growth, the service sector in India has been globally recognized for its high growth and development. The service sector has become the dominant contributor to the Indian economy accounting for more than 55 per cent of GDP and has achieved high growth with an average annual rate of 28 per cent during the last five years. The success in this sector is regarded as "India's services revolution". At the disaggregate level the structure of service sector in India is still dominated by traditional or old services, followed by business services (finance, insurance and real estate) and transport and communications. However, the new services like software services are catching up rapidly. Recently, communication services achieved the strongest growth in India. Though growing rapidly in recent years, research related services are still relatively small in India.

The role of services in Indian economy is clearly more significant. Since 1990, India's service sector has grown faster than manufacturing sector. In terms of creating new jobs, services have played a vital role in Indian economy. For example, during the last decade, India's services provided about 54 per cent of new jobs in the country. As India has to relocate a large pool of rural surplus labour, further growth in employment in the service sector will be indispensable.

India's commercial services exports accounted for almost 60 per cent of total merchandise exports in 2006-07 making the country a net exporter of services and was among the top ten exporters of commercial services in the World in 2006. This is largely due to India's success in IT service exports and relatively small value of merchandise exports.

The experience of economic development shows that, while a country expands its manufacturing capacity, the agricultural sector in the economy declines relatively, and that over time the services sector grows and eventually overtakes the manufacturing sector to become the dominant part of the economy. Information at the disaggregate level also implies that India is still at the early stage of development which is represented by the dominance of the old services and weakness in real estate, education and research services.

Growth of the service sector is determined by several factors such as production specialization, income level and urbanization. These factors are interrelated. As an economy grows, productive activities become more specialized and urbanization accelerates due to the rising level of income. As a result of increasing specialization of production, firms tend to outsource many service activities such as legal, accounting and security services. Furthermore, urbanization contributes to the growth of the service sector in two ways. Unlike farmers who to some extent can provide self-services, urban consumers rely on the market for the supply of services. They are also more likely to enter the urban informal sector for employment if there are no job opportunities in the formal sector. Thus, urban residents are both consumers and suppliers of the services. It is expected that the service sector grows as the level of urbanization increases in society.

In addition, the participation of women in the workforce has an impact on service demand as well. More women in the workforce could lead to an increase in demand for services ranging from babysitting and catering to tuition and beauty treatments. Furthermore, regulatory policies also affect the development of the service sector. A good example is the rapid growth of telecommunication services after deregulation in India. This phenomenon can be repeated in other areas such as insurance, banking, health care and so on.

In terms of the level of development of the service sector, there is considerable variation across the regions in India. Several observations about service development among regions can be made. Regional development in services follows the national trend. Delhi has India's most advanced service sector. The service sector tends to be more important among relatively more developed regions.

At the household level, Engel's law shows that, as income rises, consumption of food and durable goods in turn becomes saturated over time and demand for services such as health care, travel, communications and finance increases. Growth in income also boosts demand far away from home consumption of foods and services. There is a positive relationship between the level of service sector development and per capita GDP.

Since this growth has been boosted by globalization, the economic liberalization and deregulation in India has ensured sustained growth in services and India is benefited from WTO's General Agreement on Trade in Services (GATS). Therefore, growth prospect for services in India looks bright. But due to the impact of recent global financial crisis and global recession, the growth rate of Indian economy is now declining. The International Monetary Fund (IMF) has also projected the growth prospects for Indian economy to 5.1 per cent in 2009. This declining trend will affect the service sector adversely. So it is doubtful that the current momentum of growth in India's service sector will continue in the near future. However, this has been proved wrong by the efforts of dynamic I.T. sector in India and growth rate is around 8 per cent at present.

The consequence of such growth is that the GDP share of service sector will expand at the cost of agriculture. The dilemma for India at present is that services accounted for more than a half of GDP but employed less than 30 per cent of the country's total workforce. In addition, to some extent, recent expansion in India's service sector is boosted by the IT service exports. But, the IT sector mainly employed the educated urban youth, potentially leaving the majority of India's population further behind. Furthermore, the IT sector in India is relatively small (with a workforce under one million) and its rapid growth has not generated enough linkage with rest of the economy. India's service sector is now the dominant contributor to GDP growth but employment absorption is not very high in this sector. However, India needs industrialization even more as millions of rural workers have to be employed and as the IT sector is not large enough to significantly affect the growth of the national economy.

The service sector has gained at the cost of both the agricultural and industrial sector. No doubt, the rise in the service sector's share in GDP marks a structural shift in the Indian economy and takes it closer to the fundamentals of a developed economy. But if the service sector bypasses the industrial sector, economic growth can be distorted. The growth of service sector must be supported by proportional growth of the industrial as well as agricultural sector otherwise economic growth will not be sustainable.

2. Government's Role in Information Technology

India is the global IT superpower and a front-runner in the age of information revolution. This has been achieved partially due to facilitator role played by the Government and mainly due to drive by IT entrepreneurs in the field. It is said that political non-interference by the Government was largely responsible for such a rise in the IT field in India. The IT Department in Union Government was for a long time handled by the Prime Minister of India. Therefore, IT matters were handled most professionally. In states, the politicians did not interfere, as they were not conversant with the IT complications. By the time the politicians woke up, the IT sector has become self-sufficient and has become a major foreign revenue earner for the country providing jobs to the largest number of unemployed educated youths. None can afford now, to challenge this blue eyed boy of the country and the tiger of the Indian economy.

At this stage, Government has come to realize that it has to play a proactive role in the IT Sector. The Government of India considers IT as an agent of transformation of every face of human life, which will bring about knowledge based society in the twenty-first century. The Government of India and the IT industry are targeting for a $50 billion annual export of IT software and IT services (including IT enabled services) by 2008 over and above a large domestic IT market spread all over the country. This can only be done through a well-integrated co-ordinated and co-operative roles of Government and IT entrepreneurs. The target is being achieved as of now by the IT sector.

The Government at the Centre and the State level plays following facilitator role for the IT industry:

(a) To make the land or accommodation available to the IT industries through IT Park or IT export zones or facilitating acquisition of land under the Land Acquisition Act or purchase of land through development authorities or some other Government agencies, who have large chunks of land at their disposal,
(b) To facilitate all Government clearances like municipal licences, trade licences, mutation and conversion of land etc.,
(c) Concessional rates and sales tax, income-tax, excise, customs, service-tax etc.,
(d) To liberalize company laws in respect of IT industry,
(e) To ensure facilities of foreign travels through international flights,
(f) To allow freedom to recruit IT professionals away from normal Government procedure through employment exchange or to provide due autonomy to software company to recruit IT professional as per their need,
(g) To provide infrastructural facilities of roads, transport, accommodation etc.,
(h) To allow IT companies to work for 24 hours in violation of labour laws as they have to constantly liaison with their foreign clients who work during night in India,
(i) To make IT companies free from bandhs, strikes etc.,
(j) To equip technical colleges or institutions so that well-trained professionals and engineers are available for the IT industries locally,

(k) To help IT industry in making adequate band-width available for local, national and international connectivity.

Government can play further role at the macro level by taking following steps:

1. Accelerating the drive for setting up a World-class Information Infrastructure with an extensive spread of Fibre Optic Networks, Satcom Networks and Wireless Networks for interconnecting the Local Information Infrastructure (LII), National Informatics Infrastructure (NII) and the Global Informatics Infrastructure (GII) to ensure a fast nationwide onset of the Internet, Extra nets and Intranets.
2. Accelerating the rate of PC/set-top-box penetration in the country from the 1998 level of one per 500 to one per 50 people along with a universal access to Internet/Extra nets/ Intranets by the year 2008.
3. Policies should be drawn for setting the base for a rapid spread of IT among the citizens, propagation of IT literacy, networked Government, IT led economic development, rural penetration of IT applications, training citizens in the day-to-day use.
4. IT services like tele-banking, tele-medicine, tele-education, tele-documents transfer, tele-library, tele-info-centres, electronic commerce, public call centres or public service centres like in Hyderabad and in West Bengal.
5. Training qualitatively and quantitatively World-class IT professionals. This is worrying IT industry that the quality of new IT professionals is going down in the country.
6. An Institute of Computer Professionals should be set up on the pattern of the Institute of Chartered Accountants.

Government's Role as Regular

Government of India has promulgated the Indian Information Technology Act, 2000. Before the Act has been implemented, many drawbacks and loopholes of the Act have already surfaced. It is hoped that Government will plug these loopholes and ensure stringent measures to punish hackers and manipulators in the IT field. The

Government should also ensure that their enforcement machinery is well trained to catch the cyber criminals and their dragnet.

This is the role, which the Government is duty bound to play. No other body has the power, infrastructure and the expertise in this important aspect without which the IT sector cannot survive and prosper.

Future

After the first phase of Economic Liberalization, Shri Narayan Murthi, the Chief of Infosys, said that after that there was no going back. According to him, the IT industries now, should look ahead. Though Government's role is always unending, yet the IT industries can treat some of the problems as constraints and go ahead. After all, India is a country of one billion people and it cannot afford to give all facilities to a sector because there are other constraints at macro level. A balance has to be struck by the Government between the people's needs and the need of a particular sector. I am sure that IT Sector in India will continue to be the most loved, respected and forward looking sector. It will bring further prosperity by earning much more precious foreign currency from the level of Rs. 60,000 crore (approx.) annually. The IT Sector in India will spread its wings in other countries, which it has already started from China. I am sure that India will lead the World in IT Sector not through the multi-national companies but through the Indian companies incorporated in India by the Indians and of the Indians.

3. Government's Role in Infrastructure Sector

After the independence, the Government from the beginning took up the gigantic task of development of infrastructural facilities or infrastructural industries as per advice of Mahalanobis under the Five Year Plans. It was envisaged that the industrialists will not venture upon these sectors firstly because of large investment in this sector. Secondly, the investment may not be profitable for the nascent industries in the country. At the same time, development of those industries known as K-industries was considered essential as these were considered essential for industrial promotion in the country. The same situation also exists today in some form or the other. It is considered the responsibility or liability of Government to develop infrastructural facilities like roads, transports, railways, airports, power, telecommunication facilities, Internet communications etc.

As role of Government is shrinking day by day, it is suggested that Government-private partnership should be encouraged in this field. Even giant MNCs can invest more in infrastructure sector than many Governments. They should be allowed to invest. The Government should play a facilitator role in this regard.

Lot of measures have been taken by Government in India in this regard. Government of India has established independent Central Electricity Regulatory Commission. An Act has been passed to allow private sector entry in the power sector.

Counter guarantees have been signed for few private sector power projects. Steps are underway to restart Dhubal Power Plant in Maharashtra set up by ex-Enron of USA, which was lying idle for a long time. Mega power project investment policy is announced. Mega power projects in the public sector as well as Independent Power Producers allowed duty free import of capital equipments. There is 15 per cent price preference of the projects under public sector and deemed export benefits are for domestic bidders for projects under public as well as private sectors. New National Hydel Policy is approved. Transmission sector is accorded infrastructure status.

The country is being connected from North to South and from the West to the East through four-lane road network, which was never thought possible in the past. Airports and ports are being developed in partnership with private firms. Urban facilities are being upgraded to the international standards. Industrial units are establishing their own standby power plants. Water pollution is being controlled in the country. Even the rivers are being interconnected. The efficiency of ports is being upgraded. Even, Government machinery is being geared up to provide better and timely facilities to the clients so that they can get all facilities or licences under one roof. Economic liberalization in the first phase has tremendous effect. India has already become superpower in the field of Information Technology and telecommunication. India is shining and infrastructural facilities so created have contributed a lot to this progress.

Information Technology sector was earning about rupees sixty thousand crore from exports every year which is likely to increase substantially in coming years. There is lot of investment in real estate sector which will provide the best accommodation which can be compared to anywhere in the World. Housing is no more a problem in the country.

India has the largest percentage of young population in the World and educated manpower is available easily and at the lowest cost compared to other parts of the World except China. Agriculture sector is being liberalized to make provision for growth of industries. All villages in the country are being connected by road network under the most ambitious programmes like Prime Minister Sarak Rojgar Yojana, NAREGP for rural development. Every villager will be assured at least one hundred days employment. Similarly urban poor are also being taken care of by the Government. There is disinvestment in Public Sector and subsidy regime is slowly making the place for competitive regime. There is foreign investment in oil sector and new oil wells and gas reserves are found which will provide stability to the economic infrastructure. Tele-communication services have become cheaper and are easily available. Efficiency of coal sector has improved.

India is already a giant in research and technology. Prime Minister Shri Manmohan Singh has termed India in a speech on 21st September, 2005 as 'technology superpower'.

However, the infrastructure industry has been demanding more favours or liberalization from the Government. Some of them are as follows :

(a) Norms for the infrastructure sector be relaxed, *i.e.* minimum subscription of 90 per cent not mandatory if company discloses its plan. 100 per cent book value be allowed.

(b) Fiscal concessions for infrastructure sectors to be extended from current 10 consecutive years out of initial 12 years to 10 consecutive years out of initial 15 years.

(c) Benefits/Concessions under section 80(1A) and Clause 23(FA)(1) of Section 10 of the Income Tax Act be extended to persons operating and maintaining an infrastructure facility.

It is a fact that Government cannot allow all the demands of the infrastructure sector. The sector should also learn to stand on its own treating many of the regulations as managerial constraints and go ahead like IT Sector.

4. Government's Proactive Role in the Field of Infrastructure Sector

Government can play more proactive role by taking following steps:

1. Enhancing attractiveness of infrastructure investments with initiatives such as forcing public entities to borrow directly from the market subject to credit rating, and developing investor friendly regulatory framework.
2. Activating sources of long-term funds to finance infrastructure investments by providing fiscal incentives for investment in infrastructure and encouraging insurance, provident and pension funds to invest in infrastructure.
3. Prioritizing infrastructure spending based on estimated impact on industry and ease of commercialization.
4. Upgrading infrastructure needed for exports on a priority basis, such as ports and access roads.
5. Urgently carrying out specific power sector reforms such as eliminating cross-subsidies and supplement privatization efforts with stringent enforcement of the law to prevent theft and misappropriation.
6. Undertaking financial sector reforms focused at increasing availability and reducing the cost of capital for industry.
7. Enhancing emphasis on vocational training to improve the quality of labour resources available to manufacturing.
8. Reviewing the basis for financing of public universities/ vocational training institutes.
9. Taking specific measures aimed at increasing collaboration across industries and academic/training institutes such as providing specific incentives to encourage industry funding for research programmes and support for training institutes.
10. Encouraging creation of private vocational training centres (A model based on the success in the IT industry can be followed in this regard).
11. Encouraging industry participation in quality upgradation.
12. Simplifying administrative processes to reduce business operating costs and transactional overheads.
13. Designing procedures keeping in mind the 80 per cent who

follow the rules, rather than the 20 per cent who don't separate the process/method of 'catching the guilty' from day-to-day administration of business transactions and impose swift and strict penal action on violators, to ensure compliance.

14. Undertaking administrative simplification on a priority basis for specific areas – simplify Export and Import Transactions (Port clearance procedures, Administrative procedures for claiming export benefits and Tax Administration, Excise administration and documentation, Sales tax administration and documentation and Tax rebate claim procedures).
15. Reforming regulatory law and administration in order to ensure adequate rewards and penalties for management performance *vis-à-vis* stakeholder interests.

If these measures are implemented in Governments both at Centre and State levels, we shall be able to make India as the economic superpower in the twenty-first century.

9

Land Reforms : Essential Infrastructure for Economic Growth

In a predominantly agriculture economy, liberalization should precede implementation of land reforms. Without structural adjustments, neither agriculture nor liberalization can take off. The liberalization reforms must be drawn up carefully, taking into account the possible pitfalls specially on vulnerable sections of the society and must provide measures to protect them. The content, order and speed of liberalization must be decided within a matrix of economic and socio-political constraints.

The farming labourers are the most vulnerable sections of the society everywhere. They are not organized. Therefore, this group doesn't have the collective bargaining instrument. Some farmers' associations are there but only in Communist states like West Bengal and Kerala. Amongst the agricultural labourers, traditional poor classes like Scheduled Castes and Scheduled Tribes in India are the most vulnerable. Any reform should take into consideration the impact on them. In the World, Asia has 60 per cent of global population. Asia has great majority of World's poor next to Africa only. They mostly live in rural areas. Most rural poor are landless or have highly insecure land rights. Successful land reform measures in Japan, Taiwan, South Korea resulted in substantial economic and social benefits. There was increase in production and income up to 15 per cent after 10 years of land reforms in Taiwan. There was improved nutrition, increased consumption, better education and health, political stability, empowerment of women and democratization of lower level self-Government institutions like Panchayats in India. Table 9.1 shows how the main component of land reform, e.g. distribution of Government vested land or surplus arable land distributed is directly proportional to the economic growth of a country because Japan, Taiwan, South Korea and China are economic tigers in Asia now-a-days whereas India is lagging behind on both counts.

Table 9.1: Percentage of distributed arable land in Japan, Taiwan, South Korea, China and India

Country	*% arable land distributed*	*% landless households receiving land*
1. Japan	41%	81%
2. Taiwan	44%	100%
3. South Korea	33%	64%
4. China	90%	>90%
5. India	~5%	<28%

On the other hand, in Iran 1 per cent of rural population possessed about 60 per cent of the arable land. In Pakistan, 15 per cent of the land was shared by 65 per cent of the owners. In Philippines about 3 per cent of the land owners owned 42 per cent of land. In Columbia, 0.07 per cent land owners owned 30 per cent of the country's farm land. In Mexico, 1 per cent of the population owned 97 per cent of the land whereas 96 per cent of the population owned only 1 per cent of the land. The situation in African countries is the same. The picture in Bangladesh is more gloomy as the Pakistan dictatorial rulers enhanced ceiling limit in Bangladesh up to 120 acres to woo the middle class and big landlords. As a result, Pakistan was dismembered.

In India, it will not be an exaggeration to say that three new States in India, *i.e.* Chhattisgarh, Jharkhand and Uttarakhand were separated from erstwhile Madhya Pradesh, Bihar and UP because these regions were also neglected by their parent states from the point of view of land reforms.

The United Nation's Fourth Report (1970) says that "the redistribution of land can lead to the spread of National consciousness which is a necessary precondition for a popular National Government. On the other hand, lack of land reforms, high mechanism, capital and intensive agriculture can cause agitation."

Land reform measures aim at alleviating the rural poverty in the following manner :

(a) By distributing land among the landless by taking over excess land from large landholders.

(b) By providing security of tenure and ownership rights to tenants and sharecroppers and by regulating rent payable by them to the landowner.

(c) By protecting tribal interests in land against alienation and encroachments of non-tribals.

(d) By consolidation of landholdings and thereby bringing about rearrangements of land for better productivity.

(e) By development of public lands for providing better access to the rural poor for fuel, wood and fodder.

(f) By regulating usurious money lending and providing alternative sources of credit.

(g) By providing access to women to land and other productive assets.

(h) By protecting home sector mestead rights of the rural poor on lands owned by them and providing them with house sites for construction of residential houses.

No globalization can reach the rural areas unless land is made productive because everything centres around the land in respect of the rural poor.

They cannot take advantage of any Governmental scheme enriching them because they do not own any land. Without land they cannot establish their identity. The illiterate people can neither produce the ration card nor the voter's Identity card. Their only identity is the record-of-rights of the land owned or cultivated by them which is recognized by all the financial institutions including banks.

They cultivate the lands of other persons, but they are not recorded as cultivator on these lands. Secondly, these lands are fragmented and they will be evicted if they show their eagerness to record themselves as tenants or sharecroppers. How can unstable person contribute to the production?

Many landless poor do not have even homestead plots. They reside on lands of others specially landlords without any right or title. They face prospects of eviction if they and their families refuse to be exploited by the landowners on whose lands they reside.

Many rural poor who either own land or have been allotted land, have no means to make their land productive. Therefore, despite possessing the assets, they are not able to make use of it.

Due to indebtedness, the limited asset position of the rural poor also gets depleted which takes them further down to the poverty line.

Based on above the following factors can be worked out to firm up basis of land reforms, in a generalized way :

1. The excess land should be taken over from the big landholders. The surplus land should be distributed expeditiously and should be accompanied by timely supply of inputs and investment support. Jawahar Rozgar Yojana and National Rural Employment Guarantee Programme (100 days' work) may be resorted to assist the other beneficiaries. In areas not covered by such schemes, direct financing by Government will be of much use.
2. The ceiling laws should be amended according to the National guidelines based on the conclusion reached in the Chief Ministers' Conferences on ceilings of Agricultural Holdings and included in the Ninth Schedule of the Constitution.
3. Loopholes in existing laws should be plugged so as to secure complete tenurial and ownership rights of under-tenants and sharecroppers.
4. Priority should be given to comprehensive programmes for preparation and maintenance of land records, specially where there is no statute for preparation of such records.
5. Special attention need to be paid to tribal areas. Loopholes in laws applicable to tribals need to be plugged and administrative machinery needs to be strengthened. Cadastral survey of tribal areas should be completed where not already done.
6. All tenants including sharecroppers should be identified, their rights should be recorded and permanent heritable rights should be conferred on them before the consolidation operations are started. In the programme for consolidation of holdings, land of small holders and surplus wastelands available for distribution should be consolidated in compact blocks to facilitate the future public or Government investments for irrigation, etc. to the under-privileged.
7. Publicity requires to be intensified to familiarize the beneficiaries and even the Government servants charged with

implementation of land reform laws with the provisions of the laws.

8. The administrative set-up requires to be strengthened at different levels and should be imbued with a sense of direction and purpose.
9. The field data has shown that there are large number of sharecropper tenants who have been working with the same landowner for a number of years and are entitled to be conferred the status of occupancy tenant/owner within their respective states. All such tenants should be conferred with the status of occupancy tenant/owner with retrospective effect. Even those who have been evicted or shifted from earlier plots should be restored to the status of occupancy tenant/ownership right. As suggested earlier the recognized kisan sabha/peasant organization/voluntary, organization should be associated in such effort and their evidence on the tenancy status of an incumbent should be made admissible.
10. If the landowner does not give receipt to the non-occupancy tenant, the rent should be allowed to be deposited with the nearest revenue authority. The landowner should be made liable to criminal prosecution for refusal to grant receipts. The onus of proof should be shifted to the landowner to establish that the latter is not cultivating the land.
11. The tenants in the cultivable lands of religious institutions, trusts, muths etc. should also be conferred with ownership right in respect of those lands. Those institutions may be provided with annuity.
12. Resumption of land by landowners from tenants for self-cultivation should not be allowed except in case of physically handicapped or serving army personnel.
13. Ejection of tenants for non-payment of rent should be prohibited.
14. The small and marginal farmers who are compelled to lease out to big landowners should be linked with institutional agencies like banking, anti-poverty programmes and rural development schemes (like Jawahar Rozgar Yojana, NREGP, ITDP etc.) to make their farms viable. If they lack implements,

irrigation facilities and inputs, such facilities should be extended to them.

15. In case of dispute between the landowner and the persons claiming to be tenants/sharecroppers, the burden of proof should be on the landowner to prove the negative.
16. Sharecroppers should be legally included as tenant within the tenancy reform laws.
17. Special drives should be launched to identify informal tenants/sharecroppers in all the states who would be immediately conferred occupancy tenants/ownership status. In this drive, the local peasant organizations/agricultural labour organizations/voluntary organizations/kisan sabhas and other activist groups should be actively associated.
18. Inadequacies in the administrative organization should be removed. It should be land reform oriented.
19. The recognized local peasant organization/agricultural labour organization/voluntary organization/activist organization should be permitted to file claims for the conferment of occupancy right/ownership right to sharecropper/non-occupancy tenant before the appropriate authority.
20. Cooperative farming should be encouraged in villages on the pattern of Egypt. Panchayats may be nodal agencies to organize these. Arrangement for marketing, implementation and servicing should be available at Panchayat.
21. General awareness and conscientization camps should be organized in every state having pockets of high tenancy. Apart from above, the following are the essential items of land reforms :

    (i) Implementation of legislation relating to money-lending and abolition of money-lending.
    (ii) Updating and maintenance of land records in computerized form and implementation of legislation if any, enacted therewith.
    (iii) Political will should be created. For this landless and small and marginal farmers' representatives should be given place in local Panchayat bodies and Ministries, so that they are associated at each decision making level.

(iv) Emphasis should be given on social reforms so that the farmers use the benefits occurring on permanent basis.

(v) Panchayats should be organized and be really independent in their affairs to a large extent. They should be associated in a big way to implement land reform measures. They may be given power to dispose of petty land disputes and development of farmland also.

(vi) The poor peasants may be provided legal aid up to the level of High Court. The various stages of land litigation should be reduced. Land tribunals may be set up to reduce the pendency at the High Courts as substitutes for speedy disposal of cases.

The World Bank Study (1997) reveals that India has the largest number of rural poor as well as landless households in the World. The two are intrinsically related : landlessness – more than either caste or illiteracy – is the best indicator of rural poverty in India.

However, adopting well intended laws does not, by itself, guarantee good results. India's land reform laws have not had desired effect. Sixty-three years after independence, India's poorest still are struggling for access to rural land and land tenure security.

Land reform legislation is within States' jurisdiction. Since independence, India's States and Union Territories have employed several land reform tools including reforming tenancy imposing land ceilings, distributing Government surplus land and allocating homestead plots.

However, some innovative measures are to be implemented. One such solution is distribution of small farm plots on which the landless person can also build-up his house. These small plots up to 16 decimal of area (less than ½ bigha) should be given to them even by way of purchase by State Governments.

In Japan, the agriculture workers are given so much subsidy or incentives so that their working on farms is more lucrative than the working in industries to ensure food security in Japan. However, in spite of globalization, people have to eat food grains. Giving only full size farms require substantial amount of land and involvement of lot of public fund which may not be available with the present States. In fact, average beneficiary in India received less than 0.3 h.a. of land. Though the latter may not be sufficient for his sustenance, yet it can provide him some basis for sustenance. Rural Development Institute, Seattle, USA has suggested that even 16 decimal plot (less than ½ bigha) can save a landless person from starvation because he will grow various cash crops, vegetables apart from foodgrains on that small plot adjoining his house throughout the year with his constant labour and supervision along with his family. In Russia, 3 per cent arable land comprised in small plots yield 25-30 per cent of value of agricultural production. In China, 5 per cent of arable land had given more than 20 per cent of value of agricultural production. Similar is the experience in Indonesia and Caribbean countries. The scheme for distribution of homestead, cultivation and garden plot in India has given similar astonishing results in Karnataka, West Bengal, Andhra Pradesh and some other states where the scheme has been implemented. At least this scheme has exploded the myth of minimum size of plot which was fixed as ideal for production.

When the size of plot of 3 bigha (1 acre) was professed in 1992 as the adequate one by the writer, it was not accepted. Now, people are giving consideration for even ½ bigha plot as worth experimenting. About 17 million landless rural poor families in India cannot be allowed to suffer any more just because economic parameters are contrary to the general beliefs. The present polity cannot be stretched beyond a limit nor unlimited or required financial resources are available with popular Governments.

It is the call of the time that the rural poor are given a semblance of justice otherwise all gains of liberalization or globalization will be run over by this hungry army throwing all monuments and calculations of globalization in the dustbin.

BIBLIOGRAPHY

Agrawal, P.K., *Land Reforms in India,* M.D. Publications, New Delhi, p. 56.

Agrawal, P.K., *India's Move Towards Sustainable Development*, MD Publications, New Delhi, pp. 62-65.

Hamstad, Tim & Lokesh S.B., *Allocating Homestead Plots as Land Reforms: Analysis from West Bengal*, Rural Development Institute, USA, Year 2002.

10

Food Security in India

Food security is the top priority concern in the country. In the World 79.79 crore people are malnourished or starving out of current World population of 471.22 crore. India's population has crossed one billion mark. Population has to survive first. And then their requirements of shelter, clothing, health-hygiene or education will follow. About 15 per cent of the World's population faces a daily struggle for essentials, including 300 million Indians who survive on less than $ 1 a day.

India is a country with very high rate of growth of population. In a democratic set-up of India and due to various political-social or religious factors, the population control can be ensured to a limited extent only. Therefore, the agriculture production has to keep pace with the growing population. This is called food security which ensures not only meeting the present demand of food grains but also ensures a safe future as far as food is concerned. The death due to starvation is considered as the worst phenomenon for the modern state. It is the first and foremost duty of every state to ensure adequate food availability to its citizens. However, the belief that God has stored enough food grains before it creates its children, is as much baseless as the claim of protagonists of globalization that high speed development will fulfil all needs of the society including primary need for food.

Let us examine food security in the context of India. Almost half of India's most severe poverty is concentrated in just five states—Bihar, Orissa, Uttar Pradesh, Madhya Pradesh and Rajasthan.

The growth of agriculture sector remained higher than the growth of population between 1950-51 to 2006-07. The production of food grains has increased at an average annual rate of 2.5 per cent while the average growth rate of population during this period was 2.1 per cent. This shows that, India almost became self-sufficient in food grains.

The rate of growth of food grains production, however, decelerated to 1.2 per cent during 1990-2007, which is lower than the annual growth rate of population averaging 1.9 per cent. Therefore, the per capita availability of cereals and pulses has declined during this period. The per capita per day consumption of cereals has declined from 468 grams in 1990-91 to 412 grams in 2005-06, indicating a decline of 13 per cent during this period. The per capita per day consumption of pulses declined from 42 grams to 33 grams during the same period.

Table 10.1 shows the production of food grains and its per capita availability in India.

Table 10.1 : Production of food grains and its per capita availability in India[1]

Year	Production million Tonne		Grams per capita per day		Total Food in grams
	Cereals	Pulses	Cereals	Pulses	
1950-51	50.22	8.4	334	60.7	394.9
1980-81	129.59	10.63	417.3	37.5	454.8
2001-02	212.03	11.08	386.2	50.0	416.2
2002-03	187.57	11.31	454.3	34.9	491.2

Source: Agricultural Statistics at a glance — 2009.

During 2007, as per an estimate, net production of food grains was only 177.7 million tonne against the population of 113.65 crore reducing per capita availability of food grains at 412.8 gms per day.

India has 328.7 million hectare area. Out of this, 186 million hectare is agricultural land and net sown area revolves around 143 million hectare (Table 10.2).

Table 10.2 : Land utilization in India from 1950-1951 to 1999-2000

Year	Total area million hectare	Percentage of net sown area	Percentage of area put to non-agri-cultural uses	Net irrigated area	Percentage sown more than once
1950-51	284.39	41.8	3.3	20.85	13.14
1980-81	304.15	46.0	6.4	38.78	42.74
1999-2000	306.05	46.1	7.5	56.65	48.51

Source: Agricultural Statistics at a glance–2003.

Table 10.3 elaborates the factors to be taken into consideration while taking matrix for food security of India.

Table 10.3 : Production of foodgrains in India

(Million Tonnes)

Year	Cereals				Pulses	Total Foodgrains
	Rice	Wheat	Coarse Cereals	Total Cereals		
1950-51	20.58	6.46	15.38	42.42	8.41	50.82
1990-91	74.29	55.14	32.70	162.13	14.26	176.39
2000-01	84.98	69.68	31.08	185.74	11.07	196.81
2004-05	85.31	72.00	33.92	191.23	13.38	204.61
2008-09	99.4	77.6	38.7	215.7	14.2	229.9

Source: Economic Survey, 2008-09 and R.B.I Handbook of Statistics on the Indian Economy.

Table 10.4 : Area under Food grains Cultivation in India

(Million Hectare)

Year	Cereals				Pulses	Total Food grains
	Rice	Wheat	Coarse Cereals	Total Cereals		
1950-51	30.81	9.75	37.67	78.23	19.09	97.32
1990-91	42.69	24.17	36.32	103.17	24.66	127.84
2000-01	44.71	25.73	30.26	100.70	20.35	121.05
2003-04	42.41	26.62	30.76	99.79	24.45	124.24
2008-09	45.6	27.7	27.5	100.8	23.00	123.8

Source: Same as above.

Table 10.5 : Per Hectare Yield of food grains in India

(Kg/Hectare)

Year	Cereals				Pulses	Total Food grains
	Rice	Wheat	Coarse Cereals	Total Cereals		
1950-51	668	663	408	542	441	522
1990-91	1740	2281	900	1571	578	1380
2000-01	1901	2708	1027	1844	544	1626
2003-04	2051	2707	1228	1989	623	1707
2008-09	2177	2806	NA	NA	617	1857

* NA: Not available

It is unfortunate that land under pulse cultivation is declining and its total production is going down. As the standard of life in India is increasing day-by-day, more pulses are required. Similarly, the land under irrigation is not increasing in the Malthus ratio in which population is increasing in India. Secondly, the bad signal is the fact that utilization of irrigated land is only to the tune of about 85 per cent.

Before, fixing food security for India, variable factors of consumption cannot be determined. There is a change in per capita consumption which has increased from 1990 to 2005 in the following ratio as shown in Table 10.6.

Table 10.6: Per Capita increase in consumption (1990-2005)

	India	*China*
Cereals	1.0	0.8
Meat	1.2	2.4
Milk	1.2	3.0
Fish	1.2	2.3
Fruits	1.3	3.5
Vegetables	1.3	2.9

This cannot, however, be a ground for non-provision of adequate foodstuff as it is a natural phenomenon worldover. However, the people divert their food habits according to availability of foodstuff. For example, tribals of Andaman & Nicobar Islands live on coconut, fish and wildlife. Saints in the Himalayas live on fruits alone whereas some people have their life on vegetables alone or on meat. Thus, allied sector of meat, egg, milk and its products, sea food constitute great support system but their availability also depends on conducive environment or proper balance of land and water.

Prof V.M. Dandekar and Prof. Nilkant Rath[2] have estimated the regained diet with 2250 calories (against an average requirement of about 2825 calories as per medical opinion) as the desired minimum level of nutrition. A farmer will require at least 518 gm food grains per day to eat. Daily per capita supply of calorie in India is 2104 (1988) while in Japan 2048, USA 3666, UK 3252, Korea 2878. Wheat and rice have higher content of calorie per ounce (20.35 gms), *i.e.* 100.98 respectively against 28 and 18.23 calorie from potato and cow-milk respectively. The figure 518 gm is acceptable because the person below poverty line mainly hardworking farmer eats more of wheat or rice and less of other things. His access to other items of food is limited. Therefore, easy availability of wheat and rice should be ensured in the society.

However, taking 518 gm requirement of food grains per capita per day in India, a person will need 189 kg food grains per year. Thus, India will require 189 million tonnes of food grains. For the present population of 113.65 crore, India will require 214.8 million tonnes of food grains and for 120 crore population, India will have to pinpoint food grains requirement at 227 million tonnes.

Thus, India is almost at par with her requirement provided there is no export of foodgrains or conversion into secondary food and there is equal distribution of foodgrains which is an impossibility. Thus, the food crisis is in-built in the Indian economy and is the greatest challenge for the policy-makers, politicians, administrators, economists, agricultural scientists and others.

The strategy can be three-fold: productivity improvement, natural resource conservation and poverty eradication:

> "In USA, 368 million acres of land was idle in 2007 which is 8 per cent of the crop land in United States under a conservative programme. India should also have a reverse system of adequate production by increasing production and earmarking land, solely for agriculture which can ensure food in India which may be termed as safety net on the food front."[3]

It is suggested by few public conscious quarters that the land sown more than once which is around 50 per cent in India, should not be diverted for industries or other allied activities except for creating basic public infrastructures like road network, power supply, drinking water supply, needs for defence, natural calamities or other similar emergencies. Area under double cropping or multi- cropping should be consistently increased and production should be brought at par with the developing countries by way of organic manures and organic insecticides so that fertility of land is not destroyed in the process which was the baneful effect of the first green revolution in India.

Thus, we can come to a safety net where even 120 million ha. lands under utilization will suffice, to feed the growing population of the country. Accordingly, about 10 million hectare of land is required to support one crore population in India as a simplifier formula. The total quantum of land can be increased by converting non-agricultural land into agricultural land. It is also possible that with the growing population, production will also grow. The area of land cannot be stretched beyond a limit. All areas under non-agriculture cannot be converted under agriculture like area of deep rivers or sea and high altitude mountains. These cannot be used for agricultural purposes. Secondly, it is to be kept in mind that a portion of land will keep on becoming infertile which has to be earmarked or left free for conservation programmes.

Food security is problem worldover except in USA, Russia, Africa, Australia and South America. In Bangladesh, land per capita is only about 6/7 decimal per person. In Bangladesh three layer cultivation has started on small plots, *i.e.* on ground cash crops, above horticulture crop and top layer of timber tree. Similar is the problem in Japan or Singapore, Hong Kong and other European countries. The population has to expand vertically in multi-storied structures or horizontally in sea like in Japan for industrial and other developmental purposes. Countries like India, China, Mexico, Brazil, Egypt, Turkey, Canada are the marginal cases. These countries have to plan their expansion on land very cautiously keeping their food security in mind which is linked even with their national security. The experience of Berlin during and after the Second World War should not be repeated whereas the people had approached for one piece of bread against full basket of German currency. Thus, minimum food production or availability is to be ensured even in urban cities and towns to avoid starvation problems due to economic sanction or blockade.

Thus, India is at crucial juncture of food safety net. If it is not properly attended to, it will give rise to abrupt price rise in foodgrains like during November-December 2009 in India which was around 15 per cent and will disturb the social equilibrium apart from all calculations of planners and policy-makers. The buffer stocks with government and concessional imports of food grains cannot solve immediate problem because it involves movements of food grains which is a gigantic task. In India, the poor people in rural areas cannot be left on vagaries of market forces because the market forces have far more manipulative skills than the villagers. They are already under the cloud of vagaries of monsoon. Therefore, they have to be assured minimum food grains in their villages so that they may not starve. It is a very difficult task in the present market-driven economy. It is to be arranged apart from subsidies from Government and public distribution system through fair price shops run by the Governments. The rural public has to be given opportunity and scope to depend on their own resources and production. They have to be made self-reliant as they have the constitutional right of livelihood.

NOTES

1. Aslam, Mehmood, *Population Development and Food Security, Resource Conservation and Food Security*, Vol.II, Concept Publishing Co Pvt. Ltd., New Delhi-110 059, pp. 644-645.
2. Agrawal, P.K., *Land Reforms in India*, p. 202.
3. Tiwan, Dr. A. K. & Nair, Jeevan (2009). *Food Security and Global Economy*, Pentagan Press, 206, Peacock Lane, New Delhi-110 049, p. 32.

11

Second Green Revolution is Must to Sustain Economy

The importance of agriculture in Indian economy can be recognized by its contribution of 24 per cent to national GDP and about 65 per cent of Indian population is still dependent on it. This sector provides not only food security to the country but also employment and raw material to many industries. Foreign trade also depends on agriculture as it supplies many products for exports. Agriculture accounts for about 14.7 per cent of total export earnings. This agriculture sector forms the basis for keeping the rural people in rural areas thereby keeping a check on the speed of urbanization. The allied sectors like horticulture, animal husbandry, dairy and fisheries have importance in improving the overall economic conditions of the rural people. So the development in agriculture is an essential condition for the development of the national economy.

The average annual economic growth rate of the past decade was about 6 per cent and now the annual growth rate of GDP is about 9 per cent but the agriculture sector remains one of the slowest growing sector with annual growth rate of 1.5 per cent to 2 per cent. This is enormous burden on the rest of the economy. So this has brought to the top of the discussion among policy-makers the serious issues of reviving agriculture sector to contribute to the rural poverty alleviation and economic growth in a sustained manner.

It is imperative that the problems of farmers should be accepted with a sense of urgency. Agriculture being a state subject, the bulk of public investment in agriculture takes place at the level of States and the Central Government supports the States as a catalyst.

Agricultural products like cereals mainly rice, tea, coffee, cashew, spices, tobacco and leather are important items of India's exports and hence foreign exchange earnings. Agriculture is also the source of raw materials for agro-based industries including textiles, cigarettes, jute, sugar, paper, processed foodstuffs and vegetable oil. Moreover, agriculture sector provides market for capital goods (tractors, pump-sets and other agricultural machineries) inputs (fertilizers, insecticides) and light consumer goods.

The development of agricultural sector depends to a large extent on such core industries as power, petroleum, fertilizers and machines and tools. Thus, there is inter-dependence between agriculture and industries.

As against the commercialized nature of agricultural operations in the developed countries, farming in India still remains a source of livelihood for most farmers in India. The main reason for this is the low per capita availability of land. One important reason for this trend is fast growing population, which has adversely affected the per capita availability of land after independence. Moreover, productivity is low because traditional agricultural practices still prevail in large part of the country. Use of new agricultural technology has remained confined to few pockets of rich farmers in the selected States.

Ever since the launching of the First Five Year Plan (1951-56), agriculture sector has received the prime attention of the Government in the overall strategy of economic development. As a result, farm productivity has increased over the years and the country has achieved high degree of self-sufficiency in terms of food grains and raw materials for agro-based industries.

Per hectare yield of rice increased from 1013 kg. in 1960-61 to 2177 kg. in 2008-09 and that of wheat from 851 kg. to 2806 kg. over the same period. Total production of food grains increased from 82 million tonnes in 1960-61 to 229.9 million tonnes in 2008-09.

Although per hectare yield of major crops has increased over the last four decades, yet it is far below the international level. The green revolution did not cover pulses. The production of pulses fluctuates between 13 and 15 million tonnes per year.

Both technological and institutional factors are responsible for the low productivity of agriculture sector in India. Agriculture is still a gamble in monsoon for millions of farmers in this country, the technological factors are lack of irrigation facilities limited use of fertilizers, limited use of High Yielding Variety of seeds, inadequate plant protection measures, lack of farm mechanization, flood, soil erosion etc. The institutional factors are feudal land relations, small size of holdings, rural indebtedness, marketing difficulties etc. If the marketing system is inefficient and the farmers do not get reasonable returns for their efforts, they will be discouraged to produce more. Unfortunately, the present marketing system of agricultural produce is not favourable to the interest of the farmers. The predominance of intermediaries and the malpractices of commission agents deny remunerative prices to the farmers. Thus, low productivity in Indian agriculture is the result of a number of technological and institutional factors, which influence each other. The causes of low agricultural productivity suggest their own remedies.

A high growth rate in agriculture thus needs a comprehensive revision of agriculture policy with reorientation towards rapid diversification of this sector. A progressive correction is required in the incentive structure for agriculture. This will ensure that farmers diversify towards high value added segments of agriculture in response to the new demand structure.

The four per cent annual growth rate of agriculture production is very necessary as described in the National Agriculture Policy, but it can be turned into reality only when effective steps are taken to ensure agro-extension services and adequate and timely provision of quality inputs like irrigation, fertilizer, credit, seed, electricity to the small and marginal farmers. The present need is to enhance productivity and to ensure livelihood security in both irrigated and rainfed areas. This can be achieved if we ensure more and more public investment, empower women for their active involvement in agriculture through modern innovations and technology revolutions, sustain and expand trade in agricultural commodities.

There is no doubt that private investment in the form of tube wells, agricultural appliances, human resource development etc. does help big farmers to improve their levels of productivity. But small farmers are unable to undertake private investment effectively to boost their productivity. They have to depend more and more on public investment, which has been declining over the years. Less investment in agriculture would mean less growth of the infrastructural facilities like irrigation, rural roads, market, power, cold storage etc. and this would, in turn affect agricultural growth adversely. Development of rural infrastructure supports not only agriculture but also all rural economic activities. Moreover, in the backward states where large number of farmers are poor, public investment alone can provide big push to agriculture.

Horticulture, which includes fruits, vegetables, spices, floriculture and coconut can improve the productivity of land, generate employment, provide nutritional security and improve economic conditions of the farmers. So horticulture should be encouraged.

The farmer wants to change the cropping pattern but the real difficulty in adopting the better cropping pattern is that the farmer does not have the necessary capital to invest now or possess the know-how that may be necessary for changing the crops. Keeping in view the suicidal deaths and depression level amongst the small and marginal farmers in states like Maharashtra, Andhra Pradesh, Orissa and West Bengal, the Government has to provide insurance services to all the farmers so that in case of crop failure, natural calamities etc. they could get required financial support from the Government.

It is highly desirable that negotiations in WTO on agriculture should protect the interest of small and marginal farmers. The Indian Government is fighting for this cause and should not abandon their stand under pressure from the developed countries.

Indian agriculture faces challenges with liberalization of domestic and global markets. There is an urgent need of second green revolution, *i.e.* to develop a new strategy for the agricultural sector. Appropriate measures are required to build a productive and internationally competitive agriculture structure because promoting more rapid agriculture growth is important to lift large number of households in rural areas out of poverty and unemployment.

2. Inflation : A Challenge for Indian Agriculture

A significant rise in the general price level in a country over a long period of time is called price rise or inflation. Changes in price level are measured by means of index numbers. The price index includes a wide variety of goods and services, which are essential for life.

In India, the most commonly used general price index is called Wholesale Price Index (WPI). It is computed and released by the Central Government on weekly basis and generally taken as an indicator of the rate of inflation in the economy, 2 to 3 per cent rate of inflation is sometimes desirable for economic development. The tolerable limit of price rise or inflation depends upon the circumstances of the country at a particular time. The Indian Committee to Review the Working of the Monetary System (Sukhmoy Chakravorty Committee) recommended that inflation should not be more than 4 per cent.

In India, price stability is desired because 90 per cent of the population belongs to unorganized sector without protection of inflation. Excessive price rise causes economic deforms and social unrest and may lead to political instability. Inflation also hurts common man whose income is fixed and do not have regular jobs.

Today higher inflation rate is largely driven by fuel and food prices both of which are facing supply side pressure.

Prices of food grains act as a trend setter for general prices. Agricultural and agro-based products account for over 50 per cent weights in the Wholesale Price Index. Similarly, food and food products account for a weight of over 60 per cent in the all India Consumer Price Index for industrial workers. A sustained increase in the price of vegetables is relatively new for us. Up to eighties, the inflation has been more closely led by the prices of food grains. This change reflects the trend in purchasing power of average household and the expansion of food consumption basket previously dominated by cereals.

Fluctuations and shortfalls in the agricultural output can set off a vicious circle of general price rise. Like in other economic activities, agricultural commodities too are now faced with speculation. The speculation of future options is posing a strong negative influence on prices of traded agricultural commodities. Future trading in food grains, pulses, edible oils and sugar has introduced an additional element of price instability and marginalization of farming community, exactly the opposite of what it was supposed to check. The Parliamentary Standing Committee on Food, Consumer Affairs and Public Distribution has therefore recommended a ban on forward trading in agriculture. But only in few items, the forward trading was banned.

Forward trading in agricultural commodities has grown from Rs. 1.29 lakh crore in 2003-04 to Rs. 21.34 lakh crore in 2005-06. The power of speculation is that who has made such a heavy investment will obviously ensure that the prices do not fall. The more the hike in prices, the more is the money taken out from the pocket of poor and the middle class. Speculators then use this money in real estate and therefore the price of real estate will also rise. This should be controlled, regulated and punished.

Blaming it on the supply inadequacies is merely to divert attention from the ground realities. The fact remains that the Government has lost control over the retailers and hoarders with the dilution of

Essential Commodities Act. It does not even know as to how much of food and vegetables are actually hoarded for the sake of speculation.

In India, where a substantial proportion of the expenditure incurred by families in the low-income groups is on food grains, reasonable stability of food grains prices is vital importance. The experience in this field has shown clearly that this is a field in which, neither complete control nor complete decontrol is feasible. Government must always be in a position to regulate effectively the course of food grains prices.

The farmers must get a reasonable return. In other words, the farmer should be assured that the prices of food grains and other commodities that he produces would not be allowed to fall below reasonable prices. A policy should be made to prevent sharp fluctuations in the prices and to guarantee a certain minimum level which is essential in the interest of increasing production. Appropriate measures should be announced well in advance to ensure that the benefit accrues to the farmers. The other objective is to safeguard the interest of the consumers, and it is necessary to ensure that the prices of essential commodities such as foodgrains do not rise excessively. The key to stabilization is the building up of buffer stocks and operating on them to continuous purchase and sales over a wide front. Here, the major difficulty is that Government has inadequate storage facilities, so it is essential as a part of long-range food policy that the storage and warehousing facilities under Government's control should be rapidly expanded. Similarly, whenever or wherever, in the country prices of foodgrains tend to rise, Government should be prepared to sell adequate quantities from its stocks. These open market operations need to be undertaken flexibly and at a large number of places, so that its impact is felt directly at the points where it is needed.

The remedy also lies in increasing agricultural production and controlling fluctuations in it. This is possible through increased productivity and buffer stock operation of essential commodities. For increasing production, it is necessary to make adequate provisions for agricultural inputs and to take effective steps for institutional reforms.

Price policy in the agricultural sector struggles to achieve the twin objectives of ensuring remunerative prices to farmers and supplies to the consumers at reasonable rates. Imports of essential goods, *e.g.* vegetable oils etc. are allowed as and when necessary and exports are held back to ensure adequate supplies of essential commodities.

After 63 years, we are still facing shortage on food front. Generally, food is paid scant attention except when crisis erupts. Unless we sharply increase the production of wheat, rice, pulses and also edible oil seeds, this country would continue to face this problem.

Over a decade and a half since 1991, we can now safely assume that trade, industry and macro-economic policy reforms that have been implemented are clearly inadequate to the task. Agricultural growth needs targeted attention.

3. Role of Micro Finance

Micro-finance is a financial service of small quantity provided by financial institutions to the poor. These financial services may include savings, credit, insurance, leasing money transfer, equity transaction etc. that is any type of financial service provided to customers for meeting their normal financial needs, life-cycle economic opportunity and emergency with the only qualification that transaction value is small and customer is poor.

Micro-finance becomes distinct from other regular finances where not only credit amount is small and customers are poor, but also credit is provided with 'collateral substitute' instead of traditional collateral and non-financial services for increasing the productivity of credit.

Micro-finance is also recognized as a key strategy for addressing issues of poverty alleviation and especially women's empowerment. Through micro-financing, the transfer of financial resources to poor women enables them to become economically self-reliant, contribute directly to the well-being of their families, plays a more active role in decision-making and is able to deal with the problem of gender inequalities.

The Narsimham Committee (April 1998) also recommended that the bank policy should facilitate the evolution and growth of micro credit institutions including LABs (Local Area Banks) which focus on agriculture and any tiny and small scale industries promoted by Non-Governmental Organizations (NGOs) for meeting the banking needs of the poor. But the bankers don't want to stay in rural areas and their performance in rural branches is not monitored.

Government of India has launched a new self-employment scheme Swarnjayanti Gram Swarojgar Yojana (SGSY), which is based on group approach. This scheme replaces the earlier self-employment and allied programmes. It aims at establishing a large number of rural micro-enterprises. This scheme is conceived as a holistic programme covering all aspects of Self-Help-Groups (SHGs) and their capacity building, planning of active clusters, infrastructure build-up, technology, credit and marketing.

It envisages a greater involvement of the banks in the planning and preparation of projects, identification of activity clusters, infrastructure planning as well as capacity building and choice of activity of the SHGs.

The programme also provides for promotion of marketing of goods produced by the *swarojgaris*. Our past experience shows that the policy focus of these rural self-employment schemes has largely been on poverty alleviation rather than on creation of income and wealth on sustainable basis through investment in the productive employment generating activities. As beneficiaries of most of these schemes are poor as well as illiterates, they lack necessary skills to run effectively their activities chosen under the schemes. A big or medium enterprise can easily hire the workers with different skills.

However, it is not feasible for micro-enterprise to have access to all these skills and therefore, due to lack of necessary skills, many of them could not survive in the rapidly changing markets. Therefore, development of rural industries under group entrepreneurship may be good option in this regard.

The fruit, vegetable, dairy and meat processing industries have a great potential because in India only two per cent of fruits and vegetables are processed as compared to 70 per cent in Brazil, 30 per cent in Thailand, 78 per cent in Philippines and 80 per cent in Malaysia (as per Government of India Report (2001). And on the other hand, the middle class population has a steep rise during post-reform period. According to a report, this class has about 350-370 million population in the country and has more discretionary income to spend on quality processed food. Apart from growing middle class, urbanization, increasing of women in job market, rise in per capita income, convergence of consumers' food habits, improved information and communication technology and better transport system are other demand side drivers of growth of food processing industries. As per assessment on the basis of these demand and supply parameters, food industry in India can play a vital role in enhancing rural income and employment through agricultural diversification, commercialization and value addition. A large number of products, such as wooden and iron furniture, cotton mats, blankets, printing press, boats, candle, slate, soap, shampoo, leather products, stationery, stone work, horse and bullock carts etc. are being manufactured by micro-enterprises.

Development of group entrepreneurship, among rural youths may be good alternative strategy for rural industrialization. A group of 5-10 youths having trained in different skills can jointly run micro or small enterprises with the involvement of Panchayati Raj Institutions (PRIs) and assistance from NGOs, financial institutions and the Government.

The Self-Help Groups (SHGs) concept helps members to develop both economic and social strengths. The collateral substitute positively influences resources of the SHGs, whereby members meet their contingent obligations without going to moneylenders or private sources. The banks financing them are convinced about the collective wisdom of the groups.

The SHG-bank linkage programme has proved its efficacy in reaching and banking the unreached with lower transaction cost and better repayment performance. It is well known fact that any financial assistance, if utilized properly, generates more employment opportunities in the rural economy.

Micro-finance is productive and has a favourable effect on employment and income generation and the credit extended to rural women also has quality improving effect on the families because majority of women utilize the income generated either for investing or improving the education and health requirements of their families.

Success of Micro-finance programme has shown the way to the developing world.

These enterprises as we know, resulted in qualitative improvement of human resources. These changes, if continued in a sustained manner, would definitely bring prosperity to the village economy.

12

Work

Work is at the root of all growth and development. Without work, no person, organization or society can develop. Work is conserved. It is just possible that a society is harvesting the fruits of productive works done in the past by its members and thereby making the people of that society quite prosperous who can enjoy life. Traditionally there is gradation of 'work', due to social evolution. One work is rated higher than the other wherever all works should be treated alike. This type of discrimination is also a symptom of exploitation or capitalist society which one cannot avoid in spite of all good intentions.

According to Christ, work is the painful and glorious road that leads to God itself and the most intimate happiness. It is an instrument of virtue and a source of merit.

Work is source of income. People work in order to ensure substance, to create something useful, to improve themselves and to be in harmony with the environment or God with all its beauty, grandeur and truth.

According to 'Bhagwad Gita' a person cannot avoid work like his daily chores of activities.

The verse 5 in Chapter III reads as:

'Na hi kshchitkshana mapi jatu tisthatyakarmakrit,
Karyate hyavasah karma sarvah praktijairgunaih'.

The possession of a body and limbs binds one with the compulsion of doing some work or the other all the time, voluntary or involuntarily. The aim of action should be only to fulfil one's obligation and not satisfaction of one's desire. However, if one does work properly, both are automatically fulfilled.

None can ever remain actionless even for a moment. Therefore, helplessly, everyone is driven to action by modes of nature.

Thus, *karma*, action or work is a must for every individual. At no point of time, he can afford to be actionless. Suppose, he doesn't do positive act, he shall do the negative or harmful act. According to the third principle of law of motion of Newton, for every action there is equal and opposite reaction. He can remain actionless if external force is not applied. This is only possible in a vacuum situation.

Change is the law of Nature. Therefore, if one does not do positive action, negative action will be done on him from the nature or the World around him.

Karl Marx has also propounded the theory of continuous action and reactions resulting into incidents, which make course of society.

If a person is inactive, he is like an animal or lifeless being. He is utilizing fruits of nature without giving his due to the nature which he is duty bound. If the Sun and the Moon stop functioning in the hope of some return, then what will happen? There will be collapse of this beautiful World. Lord Krishna also says, "If you have to work as a matter of compulsion, do it happily". For this, a person has to control his 'ego' or mind which can control the senses. Only in the environment of controlled mind or senses, intellect will find its full expression. Without intellect, knowledge will not be gained. And without knowledge or the technical skill, the work cannot be accomplished properly.

Krishna further says that one should not only be expert or skilled in his work, 'Yoga is skilled work.' He should be expeditious also. For that, he uses the term 'determined intellect'.

The verse 44 of Chapter II reads as:-

'*Vyavasayatmika buddhih samadhau na vidhiyate.*'

It means that *Karmayogi* should focus on his aim fully and should not divert to aimless directions. If he concentrates on his goal, his actions will be properly and expeditiously performed. This is also propounded in the modern management.

The intellect or mind can also help to achieve harmony in *Karma* or life. Conversely, every action or work performed with the stable mind, is bound to achieve success and such infinite, when done, gradually and systematically like flowers in a garland, will result into realization of goal of the person.

A doubt may arise whether the 'Bhagwad Gita' contains formula for orderly system in society as perfect as Marx has given. The Indian culture has given emphasis on individual development. How can a person do *karma* when he finds obstacles around when he steps out of his home.

To this query, the reply is inherent in the 'Gita' which talks thoroughly to have "equity" or equality or harmony in all walks of life. It says that one should perform his duty but he should not obstruct in performance of others' duty. But one should not indulge in prohibited actions which are declared illegal by the society. He should do his prescribed actions.

Krishna instils the concept of working for a selfless goal rather than ego and ego-centric desires. If a person does self-centred activities, he gradually loses enthusiasm in his work. He develops intellectual and emotional fatigue. It ultimately results in cessation of his activities.

The worker develops a nausea for work. He even gives up his business or profession or opts for premature retirement. He who continues to work for his own interest alone, looks forward to holidays and vacations which is a clear indication of shirking work. On the contrary, a person who works for a higher ideal, a nobler cause, remains ever enthusiastic and energetic. He revels in whatever he does.

Thus, work should meet the design of life. According to Thorsrud (1972) work provides :

(a) Variety and challenge to one's ability and competence rather than activities requiring sheer endurance.
(b) Opportunity to learn and to develop one's skills and competence.
(c) Social support and recognition at workplace.
(d) Opportunity to relate what an individual performs and produces to his social life.
(e) Feeling that work would lead to a desirable future.

Work no longer remains what one does 40 hours a week for 40 years, but involves alternating periods of education, work, leisure and retirement. Thus work experience must be so integrated with the totality of life experience as to improve the quality of the whole life. Thus, labour should also be converted into work to sustain itself otherwise people will change from the primary sector to the secondary sector, *i.e.* industries and from the secondary sector to the tertiary sector, *i.e.* service sector and now the craze is to switch over to the quarternary sector consisting of financial, professional, scientific services and public administration.

This division of work is at the root of inequality in the society. Work and creation of a product are closely interlinked with the relationships within a community. Each person's work links him or her to the ongoing life of the community. In some communities in countries like India, there is close links between race, caste, groups, subgroups, sections, religion or ethnic community. In the caste system, a person is identified with a particular work. They assume their names from their occupational work. They marry among their own occupational community. Their social systems organized around the work. The occupation is the basis on which they build their lives, culture, institutions and have their work identity. In so many ways, it is the primary means of interaction and participation in society.

For example in India, the *Brahmins* are supposed to engage themselves in pursuit of knowledge and spread of knowledge. *Kshatriyas* are supposed to be fighters to rule and to fight to protect the country or society, the *Vaishyas* are to do agriculture, trade and business and the *Shudras* are to serve the society. Going by this definition the *Shudras* now known as '*dalits*' too, are supposed to engage themselves in the tertiary sector of service which is the most important sector but it is not so. Traditionally the '*Shudras*' or '*dalits*' have been exploitated and their rating of work has gone down and they are looked down by other members. Similarly, the '*Vaishyas*' who are supposed to be farmers are not farmers at all. They are financing farming indirectly and the farmers who work on the field are looked down as labourers. Similarly, the job of '*Brahmins*' has been taken over by the highly educated class of professionals, academicians or intellectuals who dominate the society irrespective of their birth or affiliations with the family. The '*Kshatriyas*' remain in the same position because they are paid wages as fighters or soldiers in army and the same is respectable. Secondly, the 'so called' politicians as '*Kshatriyas*' are the best placed section which is the elite of the society. Similarly, the work in a family is also discriminated. The hard work rendered by the women at home is not placed at the same platform as the work of men doing outside job or work.

It was believed that with the advent of the Western economic and political systems, the caste system would weaken the discrimination based on 'work'. All will be gainfully employed in work by the instrument of state as per their abilities and as per their needs. A labourer and the king will be equal. But it has remained on papers. The constitutions of various countries conferred equality on all citizens and provided equal protection of laws. The disparities have widened. In fact reverse has happened. The classes or castes which own land have acquired more land and have diversified themselves into other trades such as real estate which was the exclusive field of the 'shudra' being the service class. The weaker castes are also in those types of work which are less paying and insecure, and the majority of agricultural labourers and casual workers come from what used to be the lowest castes.

Karl Marx has exploded the myth of work and has seen intrinsic exploitation in work through the relations of production and distribution though the surplus is created as a result of work or labour. However, afterwards Karl Marx has been successfully challenged by the institutional approach which focuses on rules governing the labour relation and the relative bargaining power of individual and social classes.

The bargaining power of the labour class got further diluted because their 'so-called' proletariat leadership distanced themselves from workers and became a part and parcel of ruling class who employed them through the agency of the state power. On the other hand, the labour class avoided their responsibility. They did not work for the prescribed hours of work and did not generate that much surplus for which they could fight upon. The labourers who are engaged in agriculture, form majority section of the society in undeveloped and developing countries like India, yet they are, unorganized. Secondly, they do not want to indulge into violence as they become happy after seeing their creation in their field. It is very difficult to organize them on such vast scale. On the other hand, the state through the ballot has assumed such a huge physical or destructive power as no agitation or violent revolution can face them. It happened on the 'Red Square' of Moscow in Soviet Russia and 'Tienenman Square' of China where hundreds of protesters were just butchered by sophisticated guns, tanks and grenades along with vigilance by the helicopters.

We shall, therefore, now examine how the work can be performed without exploitation and violence and how the work still remains the singlemost important parameter to see whether a society or a country is developed or will continue to develop even in the era of globalization.

If people of a society or a country do not involve themselves in active work, that society or country is bound to decelerate in growth in long run. Initially the society or country indulging in luxuries or ostentatious living may feel very happy because they have their business interest elsewhere and have been earning from there, yet, that society or country will see its economic downfall soon when the next generation will take over. In this regard, USA and Japan have to be very careful. The generation which has been nurtured in luxuries will not have even acumen or skill even to preserve or protect the riches or wealth handed over by their predecessors. Even the maintenance of wealth or living at the same high standard requires work and capability. Sometimes, hardwork can substitute capability but the reverse is not true. It is evident that China's leap in the present day international financial field is due to hard and cheap labour contributed by its workers. The developed countries like USA, Japan, Germany, UK etc. through their multinationals are trying to exploit labour of other countries by placing their manufacturing units in developing or undeveloped countries, yet this arrangement cannot go on and on. Ultimately, the citizens of the developed countries have to work on site. Consumerism does not push up the economy *ipso facto* though some modern Economists believe that the increased purchasing power in the hands of public gives fillip to the economy. This is exactly happened in the recent downturn of economies of USA, UK, Germany, France, Japan and other developed countries. When there is height of the economy, professionals, experts, technicians and technical experts will rush to these countries to work on lesser compensation. In the regime of globalization, this cannot be stopped by regulations or restrictions of various Governments. The markets rule the roost.

John Maynard Keynes, had accepted that market is not panacea to cure all ills. In fact, Government interventions are must to save the poor workers or people at the lower ladder of the society. His prescription for employment through a capitalist system proved effective during the 'great depression' in UK There is no difference of opinions among all ideologies that every poor man should become rich and prosperous. However, difference is about the path or the means to be adopted. Marxists say that it is violent revolution which will bring prosperity whereas Keynes leaves it on market forces. The fact is that it is work or action which will bring ultimate well-being and prosperity to the individual and society. It does not mean that a worker or labour or a person in office should keep on working without being vigilant towards his exploitation, yet he should always be cautious that he should generate some surplus to protect his future and the future of his progeny.

The socialists have made workers over conscicous of their rights whereas the votaries of globalization are misguiding the masses that they should leave on market like God and they shall bring the dawn of prosperity for them. There is no denial that every individual's basic needs of food, clothing and shelter should be first fulfilled. Then there is a scope of further discussion as regards prosperity.

This is a wrong notion that morals will follow the prosperity. In fact, it is through path of morality, non-violence, toleration that society will enter into realm of egalitarianism. The egalitarian society permits violence against the exploiters or bourgeoes by way of his social or economic boycott, protests or even exemplary sanction by the state. Here, a certain amount of group violence against the exploiter is permitted. A person who is assured of justice after hard work, can really be productive to the society or the state and quality product will outflow from his labour.

Contrary to the Western approach, the Indian approach starting from the Vedas, Mahavira, Buddha to Gandhi, outlines a decentralized system of rural development and believes in emancipation of all. The Indian thinking lays down the perennial happiness which lies in spiritual attainments culminating into realization of 'self' which is also called, 'no action' requirement for the individual. Through the Indian way, satisfaction and thereby comforts are soon realized without any violence in the society. But that is not all. Individual cannot withstand against social disorder or violence. Therefore, definite system should be in place to put the polity in order or in check. In reality, the competitive economies have not created happiness but have increased mental tension and rift among people. Everyone is running for material gains without understanding what he actually aspires to achieve. The Socialist systems except in China and Vietnam have failed. Similarly, the Capitalist economies, are also on the brink of future. Some solution is to be struck in between. Excessive competitions among economies will lead to the Third World War. Therefore, an economy should be based on localized contribution of work and fruits of work should be made available on the doorstep of people. If the gains will come through market mechanism, then sometimes, the market mechanism will be so manipulated that the benefits will not reach to the real deserving persons. Somebody in between will hijack its gains. People are gradually losing faith that particular work or action is superior. Therefore, a person in a Government department or other similar sector will no longer be able to claim superior status in times to come.

For the Greek, a perfect man is a nobleman, who does not engage in labour, has leisure, takes part in war and produces spiritual work. The Romans followed the same model. They despised most work activities and acquired slaves to perform them, thus keeping themselves available for the intellectual pursuits in life. The French noblemen before the revolution considered any type of profession unworthy of their noble class. The Egypians also considered it insulting to do menial jobs. As a result Greek and Roman civilization had gone into oblivion after sometimes. The French noblemen were overthrown by the revolution. So also the Egyptian overlords were replaced by the democratically elected public representatives and popular leaders like Nasser took over. In Muslim dominated country of Africa, the Sheikhs and oil lords are behaving in a similar fashion and they still be soon thrown out of their fortresses by the new professionals of the West who are working tirelessly for them. Similarly, the corporate giants of US multinationals who are travelling in the private jets, staying in the costliest hotel suites in the World and spending lavishly on their living and leisure have been partially thrown off by their own mentor, *i.e.* globalization and will be shown door by the public in long run.

In India too, the Kings and Maharajas have been dethroned and now, they are begging for living because they hated physical labour. Similarly, the ex-zamindars and landlords can no longer forcefully take the produce on their lands because their contribution to the process of production either mentally or physically is negligible.

The political elite in undeveloped and developing countries who are cornering wealth of these nations will be soon replaced by the hard working professional-*cum*-politicians. The states like West Bengal in India, where middle class abhorred from physical work, was kept away from their say in the governance for long thirty years, by the Left Front Government who continuously ruled over the State on the basis of misdeeds and exploitation by the "*bhadra lok*" of the peasantry in West Bengal. If the party organ or cadres, do not behave properly and behave like power centres without work like in erstwhile Soviet Union, they will also be thrown off by the public. In democratic set-up, it becomes easier than the dictatorial or militia run set-up as because the public gets recurring chances as a matter of right to replace them after a certain period. However, if the proletariat class or working class does not work as per norms or as per duties, they will be thrown away by the market forces and they will be left starving for long time. Thus work or action is just like shining God which is the only refuge and is the parameter whether an individual, group, organization society or nation is going to survive in the long run of history.

The "Bhagwad Gita" therefore, propounds that one should not abandon his prescribed duty or work. There is no greater sin than this.

The action one has done, cannot be destroyed until it has borne its fruits. That is another advice that one should not bother about fruit because that is not in his hands. He should continuously work without caring for the followers of Karl Marx that he is being exploited. One has to live in the World, imbibe its influences and work as much as one can do, but not for his enjoyment sake. Work is his only companion. Work will be his only claim or capital which he can ask for. That this World is created for our enjoyment, is the most wicked idea that holds us down. In earlier days, it was easy to stage blood revolution because individuals were at the helm of exploitation. Now in the modern World, the invisible market Gods are there. There are groups or coterie or cartel behind exploitation. One cannot fight against them. Therefore, one should not be guided by the Unions or Associations alone as they work mostly on behest of their political masters or conspirators. The only way is to develop work ethics and skills so that the invisible controllers of market forces are subjugated by more physical labour at a lesser cost which succinctly helps to develop skill, intelligence and technology in the sincere worker. In Japan, workers protest only by wearing black badges and producing more. We have to follow Japanese work culture. The society which is based on discrimination of ratings of work much beyond a limit, will go into decline and the society regulated and controlled by the workforce will emerge successful. But the workers should not detest themselves from work because that is their only identity or savings to them. It will ultimately defeat manipulative market forces even in this globalization era. The engagement in activities is necessary for sustenance. It is good to be active not only for earning a livelihood but also for keeping the limbs and mind healthy.

NOTES

1. Bhatt, Ela, Jhabvala, Reena; "The Idea of Work," *EPW*, November 27, 2004, Vol. XXXIX, no.48, p. 5133.
2. Mahapragya, Acharya (2002); *Mahavir Kaa Arthashastra*, Adarsh Sahitya Sangh Prakashan, 210, Deendayal Upadhyaya Marg, New Delhi -110 001.
3. Parthasarathy, A. (2008); *Bhagvad Gita*, 1A Landsand, Mumbai-400 006.
4. Saxena, J.K. (2006); *Swami Vivekananda*, Shanti Prakashan, Bank House, Mall Road, Nainital.
5. Subha, D.M. (2004); *Bhagwad Gita for all times*, Kabir Peace Mission, 1/14 Viswas Khand, Gomti Nagar, Lucknow, 226010, UP.
6. Sinha, Jai B.P. (1980); *Work Culture in the Indian Context*, Sage Publications India Pvt. Ltd., M-32, Greater Kailash Market-I, New Delhi-110 048.

13

Imbalance in Society and its Impact on Economic Growth

The object of every system is to implement justice in the society. Justice means equality and making society egalitarian. Any imbalance in justice will create bubbles which may burst any day in the form of unrest, protests, violent demonstrations and even bloody revolution. Therefore, all systems profess to implement justice but their methods are different. We thus compare two systems from their methodology point of view. Globalization also aims at that. If imbalances exist, the economic growth cannot be sustainable. It will be an artificial, temporary and transitory growth. Thus, for globalization to succeed, the balance, *i.e.* sustainable balance should be maintained so that competition and market growth could stay long. It is also called globalization with human face. Globalization should not be like the present one resulting in urbanization undermining the traditional rural societies. It has created more poverty and gulf between the rich and the poor has widened. Many find that their jobs have gone and their lives have become more insecure. They have felt increasingly powerless against invisible forces beyond their control. The developing countries met at Doha and wanted a faster trade agenda but the imbalances of the past have not yet been redressed. One has to learn lesson from the economic crises of South-East Asia in 1997.

Joseph Stiglitz writes, "Development is not about bringing in Prada and Benetton, Rolph Lauren or Louis Vuiton for the urban rich and leaving the rural poor in their misery. Being able to buy Gucci handbags in Moscow departmental store did not mean that country had become a market economy. Development is about transforming societies, improving the lives of the poor, enabling everyone to have a chance at success and access to health care and education."[1]

The World Bank assessed that developing countries would gain 65 per cent from liberalization in tariffs which have to be more or less uniform irrespective of national borders. However, at Doha (July 2008) agricultural goods were left out of discussion fearing adverse reaction from majority of nations and majority of the people and civil society. The developed countries should not use, intensive tariffs on industrial products of export interest to developing countries in labour intensive sectors as bargaining chips.

One outcome of Doha round of discussion was good regarding rationalisation of tariff structure around the world. As a result, number of slabs of tariffs were reduced to a logical number. Similarly, earlier discussions in Cancun in 2003, the mini-ministerial conference in Geneva in 2006, the G-4 Potsdam meeting in 2007 failed because USA wanted the developing countries to open their borders including agricultural sector and liberalization of tariff on industrial products beyond their capacity at the cost of their farmers and workers in industries. In these convocations, WTO and its chief Lamy was found working for interests of USA. India walked out of the meeting. China later on agreed with India. USA could not face the cotton issue and US did not agree to change its domestic support to the agriculture. In spite of strong backing by Brazil and European Trade Commissioner and offer of about 14.5 billion as aid, developing countries like India, China, Indonesia, Kenya and many other countries did not agree to compromise with their food security and farmers' livelihood.

At Cancun Ministerial Conference of WTO held in September, 2003, a multilateral framework was sought to be made for trade and investment. The bulk of FDI flows originate from top ten industrially and technologically advanced countries led by USA for as much as 74 per cent of FDI outflows. Naturally, these countries want some sort of protection for their investment even through MNCs. But the host country has also to protect its domestic industries. The developed countries here also emphasize on their indigenous content of products to the extent from 75 per cent to 85 per cent. Is it liberalization? It is selling of products and mobilising capital of the developed countries. The unrestricted capital mobility will also destabilize the economy of the host country but they do not allow labour mobility from the undeveloped or developing countries which is very cheap as compared to their labour force. There were no indications to balance the interests of developed and developing countries. They only wanted dispute settlement mechanism in place as mandatory for protection of investment. This is possible if it is between two equals because such types of international dispute settlement mechanism is high cost intensive and tilted towards the developed countries. Judiciary is now become another forum to protect capital like the religion used to provide protection cover to the dictators, kings and arbitrary emperors.

Role of IMF, World Bank and WTO

When Thai Baht collapsed on July 2, 1997 like Lehmen Brothers collapsed in USA in late 2008, its tremors were felt from Asia to Russia and Latin America and threatened the entire World.

It badly hit Malaysia, Korea, the Philippines and Indonesia. Baht overnight fell by 25 per cent. The IMF was supposed to avert and deal with crises of this kind whereas its policies during crises were partially responsible for the downturn. It became blind to do excessively rapid financial and capital market liberalization. The World Bank also avoided studying the region, though it could give lessons for others. IMF and World Bank are obsessed with US economy and its functioning taking it as a role model. Now they must have a second look after the almost collapse of US economy and shattering of European economies during 2008 and recession and depression during first half of 2009.

The crisis deepened, unemployment soared, GDP plummated and banks collapsed. One fine morning two banks one each in UK and Germany declared themselves bankrupt and closed down. The unemployment rate was up four-fold in Korea, three-fold in Thailand, ten-fold in Indonesia. The IMF did not pursue structural reforms after providing billions of dollars as loan to these countries. Sitting in costly chambers and formulating theoretical policies of globalization was easier than to go up to the field in less developed countries to implement best practices. And when they imposed serious conditions, like hikes in interest rates, reduction in Government spending, political changes and economic changes, the economies collapsed. Thus, blind pursuit of liberalization for the sake of liberalization is not correct. It should be accompanied by the positive impact at all levels including at the grassroots level. IMF did not take up the poverty reduction as one of their agenda and left it to the World Bank. It has completely overlooked the cause of undeveloped countries. Today, market fundamentalists dominate the IMF, who believe that markets are gods and Governments do not work well. IMF works even against the theory of Keynes. They should not allow the money to go into speculators' pockets by way of playing with fiscal policies and capital markets. It is widely agreed that bailouts are no solutions which the American Treasury adopted in 2009 to bailout its major companies like Lehmen Brothers, General Motors, Citi Corps and few others. It is like printing of currency by the developing countries to avert crisis. However, bailout proves the point that ultimately nation state is the sovereign and the players of markets cannot claim that status. They do not deserve that supremacy. The core mission of IMF is to help countries to tide over their currency crises. The fund was created to deal with the liquidity crisis caused by the credit markets occasional irrationality. It refused to lend to countries that were in fact creditworthy.

All in IMF, World Bank and WTO like to sit and dictate like self-assumed Gods or policy-makers how the nation states should behave. If they do not behave in a way they want, they would impose such sanctions that their economies after being liberalized, will also collapse in few days. The political leaders and Government bureaucrats will be compelled to surrender to the dictates of mandarins sitting in these World bodies run on shares/contributions from all countries.

In a way, servants are dictating to their masters.

Imbalance at the National Level

Globalization has created centres of glaring prosperity but the deprived centres remain as they were. Special Economic Zone (SEZs) have been created all-over the world giving them concessions in taxes with a view to promote exports. They also push their finished goods illegally in the local markets creating imbalances in the local market rather than bringing in enough wealth for the country in the form of foreign currency which is the most sought after beautiful maid in the present day World. SEZs are created by acquiring lot of agriculture lands and the land acquisition evictees have raised slogans against their creation as if they were deprived of by way of payment of lower rate of their land than the market value and the subsequent higher value. They also want a share in the kitty. Except the supporters of globalization, none is supporting these SEZs which are like oasis in the desert. There cannot be two systems at one place. SEZs were supposed to be near the ports, sea or river but actually it is not so. They have shifted in the mainland. Even so many industrial parks like Information Technology Park, Auto Parts Park, Hosiery Park,

Garments Park etc. have been allowed to acquire land for the purpose at the concessional rate and they have been given lot of subsidy from the Central Government apart from many tax holidays or concessions in excise, sales tax or VAT etc. jointly by the Central and State Governments. Long established rival industrial units look them upon as their pestered competitors created by the Government. Their allegations to some extent have some basis. These units evade tax, employ less number of people than the promised ones and even indulge in flight of capital to other countries through SEZ route.

Third criticism at the National level is that the globalization has invited foreign educational institutions, training institutes, NGOs who operate on very high rates of salaries to their managers than the local ones. On the other hand, developed countries like UAE, USA, Australia have exploitative rates of taxes in their countries for the persons working from developing countries. In Dubai, UAE it is compulsory for a foreigner to spend 1/3rd of his income on rent. This is unfair. The persons working in those developed countries should get the same wages and similar facilities and privileges as their citizens get if the globalization is working in the right ernest.

The downtrodden sections like *dalits* and women in India hardly have got advantage of globalization except cooking gas for women. Though in places like the North-Eastern parts of India, the local administration is controlled by the local tribal chieftains, yet the extraction of their natural resources is continuing unabated. All foreign reserve (forex) has not come as a result of globalization. In fact $ 10 billion in 2001 came from the remittance from NRIs all over the World and they have sent back the money out of family bonds and influence of the Indian culture on them.

The Marxist, proclaimed to protect the working classes but they are only centred up to giving catchy slogans rather than working in practice. At one point of time, Karl Marx said, "I am not a Marxist." Thus, time has come when everybody has to work for the real good of masses. The globalization is paying only lip service to them and therefore, it is going to meet the disaster very soon. The globalization has partly achieved its end by exposing politicians, corrupt bureaucrats, shady union leaders and have made everybody conscious for an efficient and transparent system. But it may turn out as another slogan only. It is creating restless masses who are more dangerous than the organized masses ready for bloody revolution.

Though Information Technology, MNCs, BPOs and telecommunication giants have provided infrastructure for an efficient system, yet all the benefits are not perçolating down to the poor men. None can deny the benefits from the mobile phone which has made the management easy in the developing countries. Globalization has reached televisions and cars in remote areas, and has linked every place on this planet with global information and education through computer and e-mail. Thus, the globalization is at a crossroad. It has to further prove itself.

Imbalances at the Regional and Societal Level

Imbalances are quite pronounced at the regional, local and societal level. The provinces which have received foreign capital investments are rated better than the have-nots. They receive so much in taxes that their Budget can be supported from the receipt of taxes as a result of foreign investments. Everybody is happy from the political executive up to the man in the street because money flows for development. Problem arises when one goes deep into the long term effects of such development. The provincial Government burdens its successor Governments with loan and interest liabilities which they will not be able to discharge on their own what to say development. They then take more loans and an uncontrolled process of borrowing continues unabated to keep the politicians in power violating sound financial norms that no organization or Government should take more than 30 per cent loan of its GDP. Attempts were made to control this unethical and unhealthy practice on the part of the executive but that law is still awaiting its passage in Parliament in India for many years now and has perhaps lapsed. Thus, generations are born indebted and are fully immersed in loan. It is lucky that India has foreign reserve at the level of $263 billion (approx) as in January 2010 which can at anytime take care of its total debt burden. But the government must not be allowed to play with economy and should not be allowed to lower down the ratio of debts with GDP beyond the permissible limit. Thus, globalization can also not be hailed to have brought lot of wealth in India. The discrimination becomes more glaring when managers of MNCs and high profile or blue chip corporations or companies draw salaries to the extent of ten to twenty times of the highest salaries in the Government or elsewhere or in the indigenous private sector. The latter feel inferiority complex, though their salary bands are already more than in the ratio of 1:10, *i.e.* the highest man draws at least ten times higher salary than the lowest in the ladder. The salaries drawn by senior managers of MNCs or financial organizations have no correlation with the work rendered by them.

The service sector employees or managers draw higher salaries as compared to their counterparts in industrial sector. Thus, inter-sectoral discrepancies create heart burning among employees with the same qualification.

In the society, the industrialists and other proprietors of service sector organizations or commercial sector maintain luxurious lifestyle because as a result of globalization, personal tax structure is also liberalized to create competition. Without competition, globalization cannot survive. Some of these private management magnates, indulge in show bitz which create jealousy against them. They travel in imported cars, go on foreign jaunts, enjoy holidays in and outside the country, spend lavishly on their family functions and weddings and even maintain private jets. The young boys also roam with cells and want to enjoy life rather than sit lonely and study. Since money is there in families, this is spent on children first. As a result, children in developed countries have become obese and have become idle and dull. They have no motivation to study difficult subjects like mathematics and science because they can enjoy all facilities in student life even without their labour. As a whole, the future generation is being spoiled like in USA, UK and Japan. The Governments in USA and Japan have started worrying about it but it is too late to mend. The young boys from India and China and other developing countries have already occupied their places at a lower compensation and give more output and quality product. Naturally, employers go for workforce from the developing countries. The Ministers, *i.e.* political executives are not bothered because they require vote to remain in political power. They will get only when they can ensure all luxuries available at the doorstep of their voters at cheaper prices with ease. Everyone awakes when anti-social elements and mafia grab political power and use force and violence to achieve the economic dictates of MNCs and local business for hefty payments. The MNCs do the job by sending their managers across the globe. Thus, their job is also simplified and got done by others without any exertion on their parts. Thus, everybody seems happy in globalization. But this is the upper crust of globalization. The simmering discontent in every country will destroy the palaces built by globalization and its makers in the developed countries and more in developing countries because after all source is one, viz. public money invested through banks and other financial institutions.

However, none can deny the obvious advantages of new system. It has provided first ever solution against corruption of Government officials and their political elite. Many regulations and statutes like the Right to Information Act in India and Transparency Act in other countries have come in force to control the corruption and delay. Thus regulatory regime has assumed equal importance and more responsibility in the liberalized economic order otherwise general public will be exploited and will resent against. None imagined that the corruption could be controlled in Railway reservations in India until all reservations were computerized. Similarly, tax collection in India has gone up leaps and bounds after the computerization of system of assessment and collection of income-tax and excise duty. The public can easily access public utilities on the net and can deposit charges, fees, fines or taxes, through the computer. Everybody can talk to his business counterpart, friend or lover sitting in a foreign country through phone, e-mail as if they are sitting in a room. But these facilities are not possible to a person in remote areas of the country or globe. His expectations have grown through communication network and media but the same facilities are not available to him. The popular Democratic Governments throughout the World are trying to reach wealth to the lowest section like through NREGP in India, micro-financing in Bangladesh, yet this is not becoming a reality. The fear is if the architects of globalization themselves go bankrupt due to their inefficiency, heavy expenditure or inaction then what will happen to this system created by them. Therefore, there is need for slow but stable globalization which masses can absorb and can make it universal and eternal. However, austerity should go side by side with the globalization.

Conclusions

Trade is good but it is not good when it is controlled by global corporations and trade rules which create imbalances everywhere. Profit and growth are only two motives of large corporations. In 1994, the expansion of the GATT into a much more powerful WTO deepened the imbalance. The WTO has taken on increased powers in areas unrelated to trade like environment as a typically bureaucratic

Weberian organization does. The WTO limits Governments' ability to put controls on investment. But WTO is miserably failing to regulate what to say to control speculations by MNCs and other companies. Therefore, powers of the WTO should be scaled down. The organizations or institutions which cannot deliver, should go as per philosophy of globalization.

The National Governments are directly and periodically accountable to their stakeholders, viz. people in a democratic set-up. Therefore, their voice should have more weight. Similarly, the World Bank has more subsidized the large energy, agribusiness and infrastructure corporations than the Less Developed Countries (LDCs) at remedying the high level of their indebtedness and restore economic growth. World Bank has also used its considerable leverage to press for structural adjustment of policies of developing countries towards privatisation, deregulation, trade and investment liberalization. The World Bank was earlier propagating best practices' models around the World which has now been forgotten because World Bank now wants and has become an active agent or actor in the globalization process as a second fiddle with WTO.

Their model has been devastating for workers, the environment, equity and financial stability. The tragedy is with commons. For example, lot of fishing in the world by fishing boats has increased the price of fish in the international market because the fishing boats are increasing at higher rate than the fish are increasing in sea. Therefore, globalization is causing drainage of primary resources of the less developing countries. The undeveloped countries are totally neglected and thus stark imbalance is created in the World. Global warming and climatic changes are another serious problems which are defeating all solutions.

At the National level, protection of food security and preservation of productivity of land are more important factors. The National Government owes its first responsibility to its citizens of protection and National security. Without peace in the society, development can not take place. Similarly, democratic system will be first casualty in a foreign rule. Lot of multilateral treaties can bind National Government and its future generations unnecessarily. One generation has no right to bind its successor generation. They should have freedom to choose. A violation of trade treaty may invite wars or enforcement of treaty by sanction by even an individual nation like the USA or China. UNO or its Security Council is weak at present. This may have dangerous repercussions. When these powerful countries will fail to manage their own economies commercially, their rulers will then seek military solutions to remain in power. Therefore, globalization should be limited to bilateralism without any sanction. The developed countries have to go out to other countries as a matter of economic wisdom because their economies are not growing at a rate at which economies of less developed countries are growing. The developed countries through their MNCs will like to share this growth cake from the less developed countries. China has no fear to liberalize or close down liberalization of its economy as it is also equally militarily powerful country but countries like India should think twice before entering into trade treaties which compromise their sovereign power as an independent nation. Thus, less developed countries should exploit economic compulsions of developed countries to open out and should not go begging before them. If one goes to USA, UK or in any European country or Japan, one finds that one out of ten of their citizens or corporations are in foreign countries, and their houses or offices are closed. The globalization should create equity. The citizens and corporations from the less developed countries should also have this opportunity on equal footing. However, there is no denial of the fact that management, efficiency and transparency have to be improved considerably in less developed countries. Then all citizens of world would be able to enjoy similar facilities and privileges in their own ways around the globe.

Thus, removing imbalances injustice, inequity, starvation and poverty should remain as the main goals of globalization rather than the processes like liberalization, opening up of economies and regulatory regimes.

NOTE

1. Stiglitz, Joseph: *Globalization and its Discontents*, pp. 251-252.

14

United Arab Emirates : The Emerging Trading Partner of India

It needs to be emphasized that search, identification and penetration into the new market must be considered as crucial and strategic aspects of the export promotion measures as the step for raising internal production and exportable surplus. Different markets may be emerging with high rate of growth in income, which reflects changes in their demand and import patterns. Prudence lies in India intensifying her export effort to those markets. For this purpose, evaluation of an export strategy, which consciously keeps a close watch on different markets and changes therein, is quite necessary.

An important element in such a market oriented trade strategy, which must in fact, precede actual exports, is proper study and analysis of the market. Such a study would indicate the size, nature and future potentiality of the market as also its basic characteristics.

In the present chapter there is an attempt at making such an exploratory exercise. It surveys the trends, features, prospects and problems of India's trade to the predominantly oil producing and exporting country – United Arab Emirates.

The oil crisis and the steep rise in the oil prices and the phenomenal increase in the revenue of UAE, has indeed opened up new prospects for trade and industrial collaboration.

United Arab Emirates : Country Profile

Before independence, UAE was known as Trucial states and the territory was under British protection. In 1968, the British announced their intention of withdrawing the forces from the area by 1971. On 2nd December, 1971, six Emirates – Abu-Dhabi, Dubai, Umm-al-

Quaiwain, Fujairah and Ajman came together to form the federation of the UAE. The seventh emirate—Ras-al-Khaimah—joined the UAE in 1972. Each emirate has its own hereditary ruler who is virtually an absolute monarch in his emirate. There are no political parties or general elections. The Supreme Federal Authority is the Supreme Council of Rulers, comprising the rulers of seven emirates. The UAE is a member of the Gulf Corporation Council and of OPEC.

With an area slightly less than that of West Bengal, the UAE has a population of only 4.4 million (2007 estimate). Of this, about 80 per cent are expatriates mainly immigrants from India and Pakistan. About 70 per cent of the population lives in Abu-Dhabi and Dubai. The official language is Arabic while English is used in commerce.

Economy

The UAE is a desert country with extreme climate conditions and sparse rainfall. This makes cultivation extremely difficult. Agriculture contributes only 2.2 per cent of GDP. The Government has poured funds into this sector for reducing external dependence. Dates, the traditional crop, remain predominant in terms of area under cultivation but in recent years, vegetable production has increased. Considerable progress has been made in poultry farming.

The wealth of UAE lies in its deposits of petroleum and natural gas. Dubai has the bulk (94 per cent) of these resources. The UAE's reserves of natural gas amount to 5.7 trillion cubic metres, which is 5 per cent of the World's reserves. Mining employs only about 2 per cent of the UAE's labour force but contributes about 43 per cent of GDP.

The petroleum industry is largely controlled by Federal Government or the Governments of the individual emirate.

The economic growth of the UAE has been phenomenal since it began to exploit its oil wealth. Before this, it was an underdeveloped country with traditional occupations of fishing, pearl-dividing and trading.

On shore, petroleum was discovered in Abu-Dhabi in 1960, and commercial exploitation began in 1962. In 1966, petroleum was discovered in Dubai and production began there in 1969. Like other oil exporters, the UAE benefited tremendously from the oil price hike of 1973. Between 1972 and 1980, the economy grew at a rate of 26.2 per cent annually.

Between 1973 and 1976, the UAE used its expanded oil revenue to build-up the country's physical and social infrastructure to diversify its economy and to improve welfare services. Public sector spending rose at a rate of 72 per cent annually between 1973 and 1976 and stimulated construction and trade. However, due to monetary expansion and supply bottlenecks, inflation was high during this period. Stabilization measures were taken by Government between 1977 and 1979.

There are various deviating estimates regarding the actual growth rate of the nation's GDP. Most of the all available statistics indicate that the UAE currently has been one of the fastest growing economies in the World. In 2006, the GDP of the UAE was $168 billion with annual growth rate of 9.7 per cent and per capita income was $37000. The country has a relatively high Human Development Index (0.839 in 2004).

The UAE has developed its infrastructure to a great extent by investing its revenue in the economy. It has been a trading centre of the Arab World and aims to strengthen its commercial traditions with superior seaport, banking and tele communication facilities.

Dubai is the pre-eminent trading emirate accounting for about 70 per cent of the UAE's imports and re-exports and half of its non-oil exports.

Apart from the large number of petrochemical complexes and industries based on crude oil, the UAE also has production units for iron and steel, aluminium and cement. The emirates with smaller oil and gas reserves such as Fujairah and Umm-al-Quaiwain have also attempted to industrialize rapidly to reduce their income differences with Abu-Dhabi and Dubai.

The UAE earns about 85 per cent of its exports earnings from crude petroleum. About 43 per cent of the UAE's crude exports are absorbed by Japan. The OECD countries buy almost four-fifth of the UAE's total exports.

India's Exports to UAE

Bilateral trade between India and the UAE has expanded greatly. India's export trade with UAE has shown a phenomenal increase from Rs. 34.94 crores in 1970-71 to Rs. 152.030 crores in 1980-81. The major items of India's exports during this period were fruits and vegetables, fresh and frozen meat, rice, manufactures of metals, pearls, precious and semi-precious stones, rubber manufactures, cotton textiles, perfumes and cosmetics, iron and steel, construction materials, transport equipments, clothing and apparels, jewellery etc.

Table 14.1 : India's Exports to United Arab Emirates

(Rs. in crores)

Year	Exports
1970-71	34.94
1980-81	152.30
1985-86	263.10
1990-91	786.75
1994-95	3974.63
2004-05	33,015.10
2005-06	38,043.30
2006-07 (April-September)	28,387.52

Source: Monthly Statistics of Foreign Trade of India, Vol. I, DGcI & S, Kolkata, Various Issues.

In 1985-86, it rose to Rs. 263.10 crores and in 1990-91 it again increased to Rs. 786.75 crores. Now it became one of the main markets of India's exports. At the end of 1990, eight Indian joint ventures were in operation in the UAE. This related to aluminium, architecture and construction, agricultural products, LPG cylinders, engineering and trading activities. In 1994-95, India's export to UAE was Rs. 3974.64 crores. The main items of India's exports to UAE were readymade garments, manmade yarns, fabrics, made ups, gems & jewellery, cotton fabrics made-ups, metal manufactures, marine products, and fruits and vegetables. The other items of exports were electrical and non-electrical machinery, meat and meat preparations, basmati rice, transport equipment, plastic and linoleum products etc. India has presented a set of 31 new joint ventures to the UAE. The Videocon group has set up a US $ 5 million joint venture Videocon-Gulf Ltd. to assemble electronic goods in Dubai's Jebel Ali free trade zone.

In 2004-05, India's exports to UAE was Rs. 33015.1 crores and it rose to Rs. 38043.3 crores in 2005-06, which was 8.36 per cent of India's total exports and became the India's second largest export market after USA where, India's exports were Rs. 76166.1 crores in 2005-06, which was 16.75 per cent of India's total exports. In 2006-07 (April-September), India's exports to UAE was Rs. 28387.52 crores which is 10.09 per cent of India's total exports and maintained the India's second largest export market after USA, where during this period India exported for Rs. 43131.89 crores which is 15.34 per cent of India's total exports.

India's Imports from UAE

If we see import side of trade till 1975-76, India's import from UAE was not very high. Table 14.2 shows that in 1980-81, India's import from UAE was Rs. 349.96 crores. In 1985-86, it rose to Rs. 644.36 crores. In 1990-91, India's imports from UAE increased tremendously to Rs. 1900.34 crores and further increased to Rs. 4813.79 crores in 1994-95. The main items of India's imports from UAE were crude oil, sulphur and unroasted iron, pyres, pulps and waste paper, inorganic chemicals. Other items were pearls and precious stones, non-electrical machinery, non-ferrous metal, iron and steel, organic chemicals, artificial resins and plastic materials, transport equipments etc.

In 2004-05, India's import from UAE was Rs. 20853.17 crores and it declined to Rs. 19091.24 crores in 2005-06, which was 3.03 per cent of India's total imports. In 2006-07 (April-September), India's import from UAE has increased to Rs. 20307.79 crores, which is 5.06 per cent of India's total imports. Crude oil is the bulk import from UAE.

Table 14.2 : India's Imports from United Arab Emirates

(Rs. in crores)

Year	*Exports*
1970-71	—
1975-76	81.58
1980-81	349.96
1985-86	644.36
1990-91	1900.34
1994-95	4813.79
2004-05	20853.17
2005-06	19091.24
2006-07 (April-September)	20307.79

Source: Monthly Statistics of Foreign Trade of India, Vol.II, DGCI & S, Kolkata, Various Issues.

Scope of Trade

Due to limited requirement of oil in UAE as evidenced by small population size and limited industrial and territorial development, a very small crude oil produced is consumed locally, so most of the petroleum is exported in crude form.

Due to huge oil earnings, UAE is characterized a land of very high national income as well as per capita income by international comparison.

One of the major implications of the richness of UAE is the big size of the market and the nature and variety of goods demanded. In fact, the basic strength of UAE is its extraordinary income level which acts as a great pulling factor for other countries to cater to this market with full consideration and care and to make this a focus of her trade policy. High degree of quality consciousness and level of sophistication, variety in demand and desire to get latest of everything, are some of the reflections of this richness which each exporting country must take notice of.

The small population base has accounted the problem of labour shortage which is a strong implication for the size of market. Abundance of financial surpluses alone could not help in the absence of complementary factors like labour and management. Hence the labour import was used as last resort. The demand for alien labour force has been there due to the fact that native labour force was neither available in required number nor did it possess even the minimum skills needed for its proper utilization. The mounting oil revenues prompted the creation of minimum basic infrastructure, housing, education etc. Therefore, UAE has to liberalize the immigration of foreign labour.

Shortage of local labour has important implication for India. India has been able to export large scale manpower to UAE. Remittances through these migrant workers have swelled their foreign exchange reserves. In India, commodity exports along with manpower exports give a full picture of foreign exchange earnings from UAE.

The prevalence of vast desert, arid lands and little rain fall have led to scantly agricultural development in UAE. With rising population and per capita income on the one hand and limited agricultural potential on the other, the food imports have become quite enormous.

India's role as an exporter of merchandise goods as well as manpower for both of which UAE is crucially dependent on others. Skilled and semi-skilled workers, technicians, engineers, architects, teachers, doctors and other professionals crossed the Arabian Sea to land up in the UAE, bubbling with oil wealth and financial abundance.

Apart from the above angles, India looks at UAE with two motives, one as a region where she has a good scope for project exports, joint ventures etc. and second where from India continues to get investment funds in the form of direct investments, loans, aid etc. to meet her balance of payments deficits.

India is in a position to offer not only some of the agricultural products but is also in a position to offer many of the manufactured items both consumer goods and industrial goods which UAE needs and the skills and expertise that they surely lack.

Cement, rice, sugar, spices, tea and textiles have been India's traditional exports to the UAE. Quite apart from the traditional items, this country offers very valuable markets in the area of machinery and equipment such as air-conditioning equipment because UAE is extremely hot. There is great scope for export of air-conditioning plants and machinery, cold storage, refrigerators, cooling plants, ice factories, water coolers, diesel engines, aluminium conductors, transmission towers, power cables, furniture items, and plants for flour milling, oil extraction, soap making, fertilizer and chemical plants, textile mill machinery, railway rolling stock, bicycles, trucks and commercial vehicles, military soft-ware, steel tubes and pipes, machine tools, power generators, earth-moving equipments, construction materials, building hardware, steel and rolled steel items, processed fruits, pickles and consumer goods of all descriptions. Similarly, very large projects of modernization, construction of air ports, roads and buildings of housing colonies, quarters and industrial plants are going on in UAE and she could participate in joint ventures and turnkey projects.

It is also very important that the quality of goods exported to UAE conforms to the highest international standard. As this country is today overflowing with large amount of foreign exchange, it can shop anywhere in the World. India's ability to introduce her products and to nurture this over a long period of time would depend upon her adherence of delivery schedules and her prices being internationally competitive. Nowadays, India's shipping services have considerably improved. Refrigerated cargo space is now becoming available on some of the vessels plying on the route. This provision makes it possible for India to export perishable items and other produce. It is also very important to initiate air charter services to this area, so that her consignments could be lifted at short notice and ferried across.

In UAE, there are three types of India's projects civil construction projects, turnkey projects and services and consulting projects.

In civil construction projects, mainly are water and sewerage projects, bridge projects, building and housing projects, irrigation and land reclamation projects, railway construction projects and other including pipelines laying projects.

Many public sector undertakings, like Rail India Technical & Economic Services (RITES), Engineering Project India (EPI), National Projects Construction Corporation (NPCC), Indian Railway Construction Company (IRCON), Tele-Communication Consultant India Ltd. (TCIL) etc. are involved in most of such projects. Many of such projects were won in open competition from firms belonging to even developed countries. The construction work of Ruwais, Abu Dhabi was major Indian project.

Joint ventures also played an important role in generating engineering exports to UAE. According to a report, "The recent phenomenal rise of exports for engineering goods, more particularly of capital goods, owes a great deal to the establishment of joint ventures and other projects in UAE. The joint ventures not only help to build-up the industrial image of India but, through their successful operation, demonstrate effectively the quality of Indian products and services. In fact, it is suggested that joint ventures should be viewed as integral aspects of India's export promotion strategy, because India has developed a sound technological infrastructure and is in a position to export "appropriate" or "intermediate" technology, which is quite suitable for most of the developing countries. The export of related goods seems quite feasible from India.

From the investment angle, India's majority of joint ventures (in operation and under implementation) are related to manufacturing activity, like light engineering, chemicals and pharmaceuticals, textiles, leather and rubber products and iron and steel products.

India's economic reform and liberalization policies on the whole have a positive impact and encouraged greater private sector exchanges with UAE. This has also been a diversification in the items of trade.

The share of eleven major trading partners of India, accounting for nearly a half of India's trade, has not changed much since 2000-01. Table 14.3 shows that US continues to be the single largest trading partner of India, but with a declining trend, China on the other hand,

Table 14.3 : India's Major Trading Partners
[Percentage share in total trade (Exports + Imports)]

Country	*2000-01*	*2001-02*	*2002-03*	*2003-04*	*2004-05*	*2005-06*	*2006-07 (April-October)*
USA	13.0	12.2	13.4	11.6	10.7	10.6	9.7
UK	5.7	5.0	4.6	4.4	3.7	3.6	3.1
Belgium	4.6	4.4	4.7	4.1	3.6	3.0	2.4
Germany	3.9	4.0	4.0	3.8	3.5	3.8	3.7
Japan	3.8	3.8	3.2	3.1	2.8	2.6	2.3
Switzerland	3.8	3.4	2.4	2.6	3.3	2.8	3.2
Hong Kong	3.7	3.2	3.1	3.3	2.8	2.7	2.3
UAE	3.4	3.6	3.8	5.1	6.2	5.1	6.9
China	2.5	3.1	4.2	4.9	6.5	7.0	7.7
Singapore	2.5	2.4	2.5	3.0	3.4	3.5	4.0
Malaysia	1.9	2.0	1.9	2.1	1.7	1.4	2.2
Total	48.6	47.2	47.9	48.1	48.2	46.1	47.5

Source: DGCI & S, Kolkata.

has increased its share almost three fold to become the second largest trading partner of India. With rising POL prices and India not only importing curde oil from but also exporting refined POL products to the United Arab Emirates (UAE), UAE has emerged as the third largest trading partner of India. While in 2003-04, United Arab Emirates was the second largest trading partner of India, after the US, contributed strong growth in exports. India's multifaceted relations with the UAE continued to grow during the year 2006-07.

Since, UAE is implementing various schemes for developing agriculture, transport, communication and other sectors, all of which would call for a substantial increase in imports of capital equipments, raw materials and consumption goods year after year, India should take advantage of these opportunities for stepping up her exports.

The need is to establish appropriate communication channels to inform and influence the internal trading enterprises of UAE for new product information. Indian exporters should seriously think of making more effective use of marketing services offered by representation agencies.

It is through participation in trade fairs and exhibition that extensive direct contacts can be established with the end users in industrial and consumer goods, officials of testing laboratories and foreign trade organizations as well as representatives of various facilitating agencies.

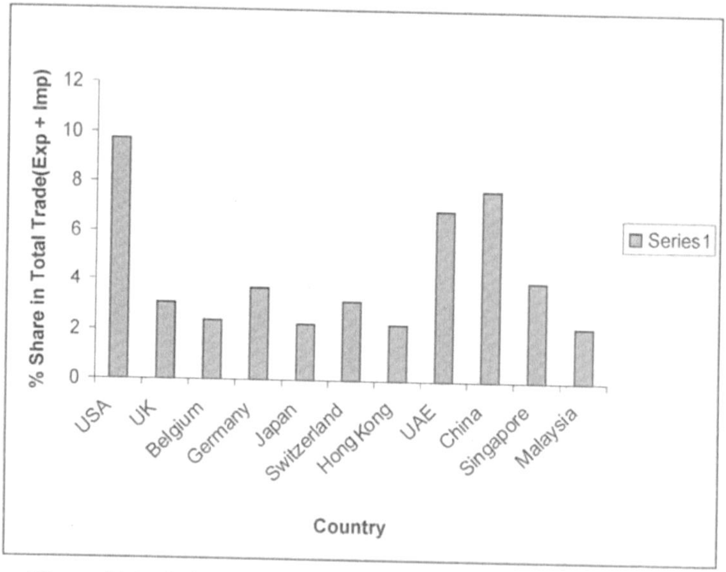

Figure 14.1 : India's Major Trading Partners 2006-07 (April-October)

Since, India may be trying to sell engineering goods in increasing volume in the future, it would be desirable to convene technical seminars periodically in selected centres for customer information and initial introduction of product. At the same time India should be beware of the autocratic functioning of polity in UAE and should safeguard investors from satiation of goods and services in UAE by placing to third country settlement of disputes under Private International Law.

Market information can be monitored on a regular basis for wider dissemination through Export Promotion Council and other export promotion agencies to the interested Indian suppliers. The product literatures and catalogues of individual manufactures should indicate maximum detail in regard to specification and end users.

Proper exchange of market information will safeguard the new investors from India who suffered during sudden recession in Dubai in December 2009.

Conclusion

One has to leave a room for consideration that the economy of UAE is largely immigrant based. Therefore, it is the inseparable part of the global economy.

This chapter concludes that the market potential in UAE is quite substantial. It is possible to improve India's market share. For this, India should first try to project herself as an important producer of selected range of commodities as far as quality, costs and prices are concerned and create an image of a reliable exporter of these commodities. Based on this good image, India can further penetrate into the market by introducing new commodities to UAE and get a further share in cake. Secondly, India should build up an emergency exit route from the high volume economy of Dubai, UAE which collapsed in December 2009.

REFERENCES

Agarwal, P.N. (1978); *India's Exports Strategy*, Vikas Publishing House Private Ltd., New Delhi.

Bhagwati, J.N. & Padma Desai (1970); *India : Planning for Industrialization and Trade Policies since 1951*, Oxford University Press, London.

Bhatt, R.K. (1999); *India's Trade Relations with the Middle East*, Saraswati Prakashan Mandir, Allahabad.

Narula, Subhash (1988); *Gulf Economies in Indian Perspective*, Common Wealth Publications, New Delhi.

India's Gulf Exports, Anupama Publications, New Delhi.

Annual Report (Various issues); Ministry of External Affairs, Government of India, New Delhi.

Economic Survey (Various issues); Ministry of Finance, Government of India, New Delhi.

Monthly Statistics of Foreign Trade of India, Volumes I & II (Various issues), Director General of Commercial Intelligence & Statistics (DGCI & S), Ministry of Commerce, Government of India, Kolkata.

Middle East Economic Digest (Report).

The Economic Intelligence Unit (EIU), London, UK, Country Report, United Arab Emirates.

15

India's Trade with China : Trends and Patterns

Economic ties between India and China are rapidly emerging as one of the most important trade relationship in the World. The World's oldest civilization, India and China today stand among the major countries, which drive the global economy. Both countries have experienced a long colonization and today bring up about for 40 per cent of the World's population. Rapid economic growth, large foreign direct investment and growing trade with all countries across the World are the special characteristics of both countries.

There is a scope for economic cooperation between India and China. Both the countries are growing fast. They are home to the World's largest pools of skilled workforce and are expected to be the engines of global economic growth in present century. Both countries are developing closer economic relations with each other and with the rest of the Asian countries through bilateral and regional agreements.

In recent years, bilateral trade and investment between India and China are growing fast, indicating the existence of a vast potential for economic cooperation.

China's joining of WTO coincided with the rapid rise in the Indo-China trade. During the recent year (2007-08), the total trade grew by 47.47 per cent. India's exports to China increased by 30.73 per cent and imports from China enhanced by 55.42 per cent in 2007-08. However, India's share in China's global imports remained just one per cent. China's share in India's global imports reached 5 per cent. In recent years China is gaining importance as destination for India's exports. Now China is emerging as the second largest trading partner for India, next only to the U.S. and also India is emerging as the 11th important trade partner for China. Both the countries need to accelerate the development to expand their economic base. India though lags behind China in many respects, has a potential to achieve the level of China in near future. Both countries reached US $20 billion bilateral trade in 2007 and target US $40 billion bilateral trade in 2010.

In this chapter, an attempt is made to analyze the trends and patterns of India's trade with China since 1991. The chapter has covered following aspects of India-China trade — first historical background of Indo-China trade, second India's trade with China: trends and patterns since 1991 and third is conclusion and suggestions.

Historical Background

Most of the 20th century marked a cold relation between India and China as they fought a couple of wars due to border disputes. The British India had been claiming the McMohan line as the line of control as agreed by the Simla Conference of 1914 signed by the British India and the Tibet. However, with the independence of India in 1947 and the formation of Peoples' Republic of China in 1949, a period of harmonious relations continued throughout the 1950s. India and China established diplomatic relations in 1950. India was the second country to establish diplomatic relations with China among the non-Socialist countries.

The doctrine of "Panchsheel" for peaceful co-existence was jointly proposed by both Governments in the year 1954 for mutually acceptable resolution of border disputes and contributes for the regional stability. Both India and China had travelled on a similar growth trajectory between 1952 and 1978, as their economic growth rate was almost the same, *i.e.* around 3.8 per cent per annum for India and 4 per cent per annum for China. However, India has a fast growing service sector whereas Chinese growth is mainly based on manufacturing sector.

India and China signed a trade agreement conferring Most Favoured Nation (MFN) status in 1984. The relation did not really improve that much during this era. However, Indian Prime Minister Rajiv Gandhi's visit to China in 1988 was major breakthrough and turning point in the bilateral relations. During that visit many agreements were signed on science and technology, civil aviation etc. A series of diplomatic visit enhanced the bilateral relations since 1988. The visit of Indian Prime Minister, Atal Bihari Vajpayee in 2003 marked a new milestone in the relationship between the two countries. The visit of Chinese President. Hu Jintao in 2006 marked great leap in the relations between these two countries. Out of all, the economic relations took the major step forward with an agreement to boost bilateral trade to US $ 40 billion by the year 2010.

There have been two major differences other than political background, which play a crucial role in deciding the economic environment of both countries — First, the approach towards economic development adopted by each country and Second, the exchange rate arrangements opted for by both the countries.

India's Trade with China : Trends and Patterns

Indo-China trade has been growing very rapidly since mid-1990s. In 1990-91, India's export to China was $ 18.2 million, which grew to $ 10834.08 million in 2007-08 registering an annual exponential growth of 31.6 per cent per annum. India's import from China has shown a similar trend. India's import from China was $ 31.0 million in 1990-91 increased to $ 27115.64 million in 2007-08, showing an annual exponential growth of 34.9 per cent. Total trade between India and China touched $ 37949.72 million in 2007-08. Total trade between two countries has been growing at an annual exponential rate of 29.93 per cent between 1990-91 and 2007-08, which is much higher than the rate of growth of India's overall trade during this period. Even India's exports to China have been growing much faster rate than its total trade. India's exports to China was $ 10834.08 million in 2007-08, registering a growth of 30.73 per cent as compared to 2006-07, when it registered growth of 22.61 per cent over the previous year. This shows that India's exports to China have been growing very rapidly and significantly but not to the same extent as China's exports to India.

India's imports from China have also been increasing significantly over last ten years. India's import from China was $ 27115.64 million in 2007-08 showing a growth over the previous year. Another significant trend is that trade deficit has also been increasing over the years, which stood at $ 16281.3 million in 2007-08 as compared to $ 9159.53 million of the previous year. Trade deficit was $ +27.3 million in 1991-92. As regards to India's total trade (both exports and imports) to China, it was $ 37949.72 million in 2007-08, registering a growth of 47.47 per cent over the previous year 2006-07. Another significant trend is that share of India's trade with China in India's total trade has been increasing continuously over the years. In 1990-91, share of India's trade with China was only 0.12 per cent of India's total trade, which grew to 9.15 per cent in 2007-08 (Table 15.1).

But studies related to trade intensity indices reflect that India and China are not trading at a level as high as it should be. Thus there is a large potential for growth in bilateral trade between these two countries.

India's exports of principal commodities to China on a time series basis from 1998-99 to 2004-05 are shown in Table 15.2. Among them, iron and ore are the single largest items of India's exports to China in 2004-05, which is 39 per cent of India's total export to China, as compared to 21 per cent in1998-99 though the trend has been increasing over the years. Primary and semi-finished iron and steel are the second largest items and their share is 9 per cent of India's total export to China. Plastics, linoleum products, pharmaceuticals and fine chemicals, machinery and instruments, non-ferrous metals, marine products, cotton yarn etc. are other major products of India's export basket to China. The exports of gems and jewellery have gone up to $18.45 million in 2004-05 from $ 1.27 million in 1998-99. Export of commodities like manufactures of metals, marine products and primary and semi-finished iron and steel have also registered increasing trend during this period.

Table 15.3 shows India's import of principal commodities from China from 1998-99 to 2004-05. India's most important item of import from China is electronic goods which is 31 per cent of India's total import from China in 2004-05. India imports substantial amount of coal, coke and lubricant from China, which was only for $ 221.5 million in 2003-04 and suddenly shot up to $ 780.19 million in

Table 15.1 : India's Trade with China

(value in US $ million)

Year	India's Exports to China	Rate of Growth (%)	India's Imports from China	Rate of Growth (%)	Trade Balance	Total Trade with China	Growth of Total Trade with China (%)	India's Total Trade	Share in India's Total Trade (%)
1990-91	18.2		31.0		-12.8	49.2		42217.7	0.12
1991-92	48.2	164.8	20.9	-32.58	+27.3	69.1	40.45	37275.9	0.19
1992-93	141.3	193.15	126.0	502.87	+15.3	267.3	286.83	40418.8	0.66
1993-94	279.1	97.52	302.0	139.68	-22.9	581.1	117.4	45544.5	1.28
1994-95	254.2	-8.9	760.8	151.92	-506.6	1015	74.67	54984.9	1.85
1995-96	332.7	30.9	812.0	6.73	-479.3	1144.7	12.78	68470.2	1.67
1996-97	614.8	84.79	756.9	-6.78	-142.1	1371.7	19.83	72602.1	1.89
1997-98	718.1	16.8	1119.3	47.88	-401.2	1837.4	33.95	76490.9	2.4
1998-99	427.2	-40.50	1095.7	-2.11	-668.5	1522.9	-17.12	75607.4	2.01
1999-2000	538.8	26.12	1286.7	17.43	-747.9	1825.5	19.87	86493.1	2.11
2000-01	831.3	54.3	1502.2	16.75	-670.9	2333.5	27.83	95096.8	2.45
2001-02	952.0	14.51	2036.4	35.56	-1084.4	2988.4	28.06	95240	3.14
2002-03	1975.5	107.51	2792.0	37.10	-816.5	4767.5	59.53	114131.5	4.18
2003-04	2955.1	49.58	4053.2	45.17	-1098.1	7008.3	47	141991.7	4.94
2004-05	5615.9	90.04	7098.0	75.12	-1482.1	12713.9	81.41	195053.3	6.52
2005-06	6759.1	20.36	10868.0	53.11	-4108.9	17627.1	38.64	252256.2	6.99
2006-07	8287.48	22.61	17447.01	60.53	-9159.53	25734.49	45.99	311866.78	8.25
2007-08	10834.08	30.73	27115.64	55.42	-16281.56	37949.72	47.47	414546.15	9.15

Source: Monthly Statistics of Foreign Trade of India – DGCI & S, Kolkata, Government of India.

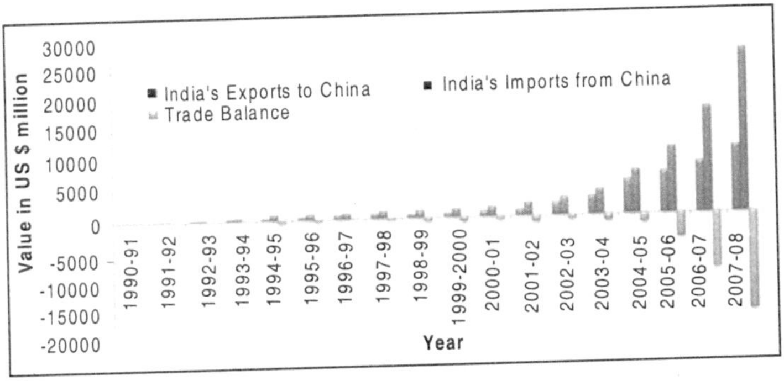

Figure 15. 1 : India's Trade with China

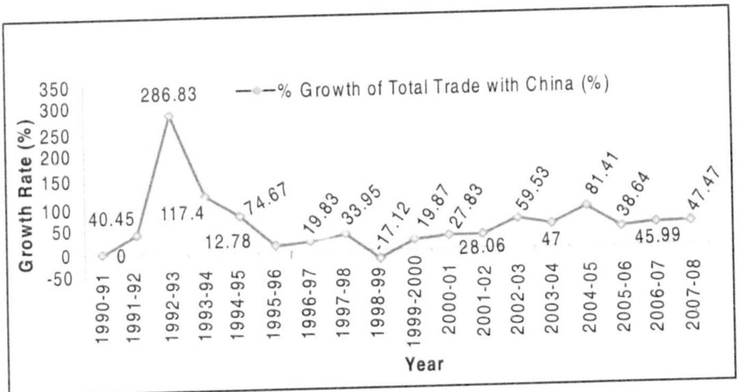

Figure 15.2 : Growth of Total Trade with China (%)

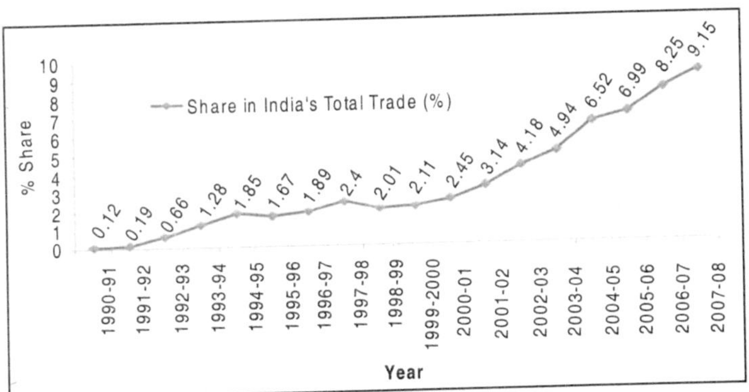

Figure 15.3 : Share in India's Total Trade (%)

Table 15.2 : India's Export of Principal Commodities to China

(Value in US $ millions)

Name of Commodity	*1998-99*	*1999-2000*	*2000-01*	*2001-02*	*2002-03*	*2003-04*	*2004-05*
Iron ore	87.77	81.43	130.16	207.56	409.37	825.74	2084.43
Primary and semi-finished iron & steel	4.72	3.59	26.4	11.02	491.02	577.25	488.74
Plastic & linoleum products	6.28	25.36	103.11	141.95	177.92	306.44	438.9
Processed minerals	16.99	11.56	15.24	23.4	87.58	99.99	230.85
Inorg./org./agro chemicals	22.28	39.32	60.01	48.84	88.89	83.96	217.91
Other ores & minerals	39.93	65.47	73.15	113.81	154.3	128.88	192.48
Drugs, pharma. & fine chemicals	42.68	46.01	58.56	80.37	93.02	102.53	106.33
Machinery & instruments	9.17	11.72	19.72	11.36	32.47	73.5	98.14
Residual chem. & allied products	6.42	8.5	11	16.75	21.45	33.89	76.74
Non-ferrous metals	0.1	0.08	17.99	2.85	21.2	43.53	65.42
Marine products	51.51	87.81	116	85.19	118.39	89.08	65.38
Cotton yarn fab. Made-ups etc.	40.73	56.05	71.54	75.06	64.04	74.11	64.92
Electronic goods	4.08	9.24	20.66	15.9	22.82	42.31	44.53
Castor oil	17.19	31.95	13.83	7.08	1.64	8.93	40.89
Oil meals	32.64	9.69	8.29	4.22	3.67	17.28	35.81
Finished leather	3.32	4.02	8.41	12.7	15.84	21.65	30.5
Manufactures of metals	3.79	1.41	10.57	11.5	32.31	27.9	28.95
Ferro alloys	2	0.54	0.15	4.54	3.93	6.2	28.44
Dyes, intermediaries etc.	3.41	3.31	8.59	11.29	20.08	66.2	18.63
Gems & jewellery	1.27	0.31	0.17	0.66	2.08	9.73	18.45
All commodities	427.06	539.41	830.3	955.19	1980.61	2962.92	5344.88

Source: Monthly Statistics of Foreign Trade of India – DGCI & S, Kolkata, Government of India.

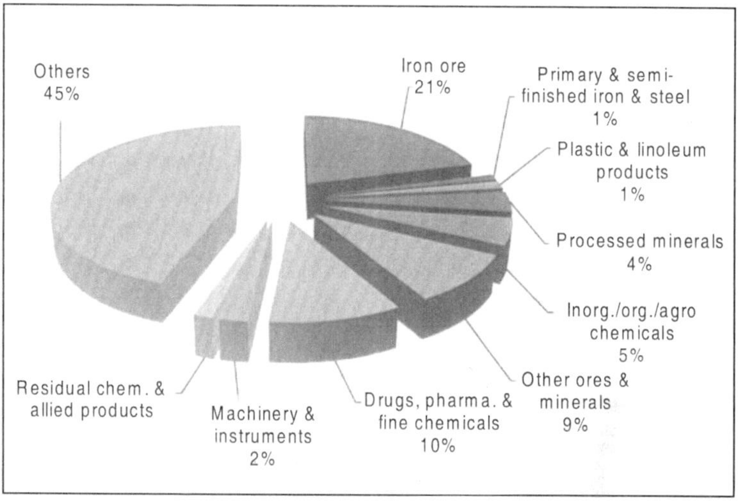

Figure 15.4 : India's Export of Principal Commodities to China (%) (1998-99)

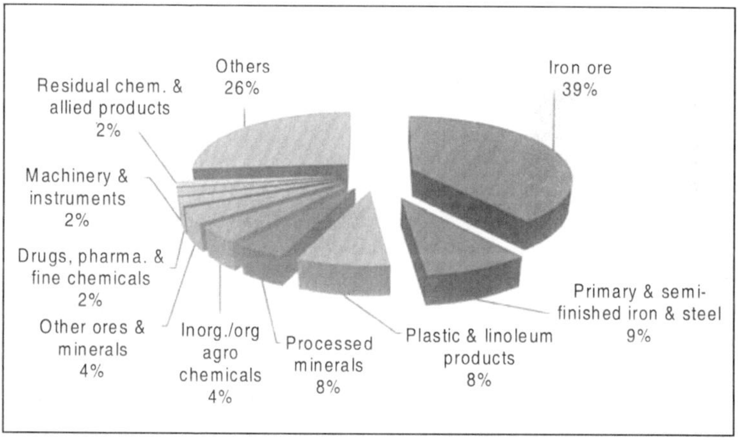

Figure 15.5 : India's Export of Principal Commodities to China (%) (1998-05)

2004-05 and become the second largest imported items with the share of 11 per cent of India's total import from China, though it was modest in the early years. Organic chemicals are the third largest import item from China in last several years. Electrical and non-electrical machinery, medical and pharmaceutical products, textile yarn, fabrics and made-ups, silk yarn and fabrics, non-ferrous metals, silver, iron and steel inorganic chemicals, raw silk, non-metallic

Table 15.3 : India's Import of Principal Commodities from China

(Value in US $ millions)

Name of Commodity	1998-99	1999-2000	2000-01	2001-02	2002-03	2003-04	2004-05
Electronic goods	161.46	178.22	244.56	385.19	812.16	1384.44	2069.19
Coal, coke & lubricants	108.2	146.04	261.34	263.43	175.61	221.5	780.19
Organic chemicals	163.54	178.71	181.02	242.38	326.42	474.06	606.66
Non-electrical machinery	48.92	53.04	60.49	72.14	105.69	184.38	424.43
Electrical machinery	20.6	28.58	46.34	59.98	72.3	101.7	211.23
Medical & pharma products	71.27	73.05	70.32	105.26	150.25	185.26	192.8
Other textile yarn, fabrics made-ups	—	20.5	34.75	46.89	75.97	107.66	172.95
Silk yarn fabrics	—	12.81	17.27	32.2	52.23	105.2	156.52
Non-ferrous metals	52.27	49.01	41.89	54.53	46.23	86.82	145.13
Silver	—	20.16	8.3	94.41	121.21	58.42	138.62
Iron & steel	20.44	30.75	7.53	14.52	9.24	31.65	136.21
Inorganic chemicals	62.72	53.2	48.37	59.66	69.25	104.46	131.22
Silk raw	48.31	87.44	95.66	122.9	106.96	113.7	123.34
Non-metallic mineral mnfs	7.42	10.24	14.65	30.24	44.42	64.58	121.12
Manmade filament/spun yarn/waste	—	12.38	9.84	17.32	56.05	82.61	115.78
Metaliferrous ores & metal scrap	8.5	11.26	19.65	14.52	28.46	63.96	102.99
Professional instruments, optical goods etc.	18.85	23.37	23.38	55.31	80.49	90.01	99.32
Transport equipment	3.49	8.45	6.62	9.06	22.15	13.16	88.5
Manufactures of metals	12.04	15.38	17.77	31.14	32.69	55.34	88.07
All commodities	1096.47	1288.27	1494.92	2043.33	2799.29	4063.96	6768.92

Source: Monthly Statistics of Foreign Trade of India – DGCI & S, Kolkata, Government of India.

manufactures, manmade filament/spun yarn/waste, metal ferrous ores and metal scrap etc. are other important items of imports from China have been identified, which are consistently increasing over the years. The share of electronic goods in total import from China is 31 per cent in 2004-05 as compared to 15 per cent of total import from China in 1998-99. The share of import of other commodities have also increased during this period (Figures 15.6 & 15.7).

Conclusion and Suggestions

India and China are now emerging as major countries not only in the Asia but also in the global economy. Chinese President Hu Jintao said during his visit to India in November 2006: "Asia will dominate next century if China and India can strengthen their trade and business links." The economic activities in these countries have become increasingly significant at the global level. Both the countries need to accelerate the development to expand their economic base. India though lags behind China in many respects, has a potential to achieve the level of China in near future. Both the countries are making efforts to improve their trade between them. Recently, India and China have decided to set up a working group that will prepare a framework to resolve the decade old border dispute.

India and China are the economic powers of the Asian region. Both the counries are growing at a faster rate. China has a dominant manufacturing sector while India has a strong service sector. The two economies can offer each other valuable experience. China can learn from India to develop a strong service sector and in terms of manufacturing development, China may offer a model for India. The bilateral trade between India and China is increasing at rapid growth rate and Indo-China two-way trade has crossed $ 37.9 billion in 2007-08. India's export to China $ 10.8 billion in 2007-08 and import from China reaching $ 27.1 billion. Both the countries are considering the conclusion of bilateral trade agreement to enhance trade. China has poised to become India's large trade partner in the coming years.

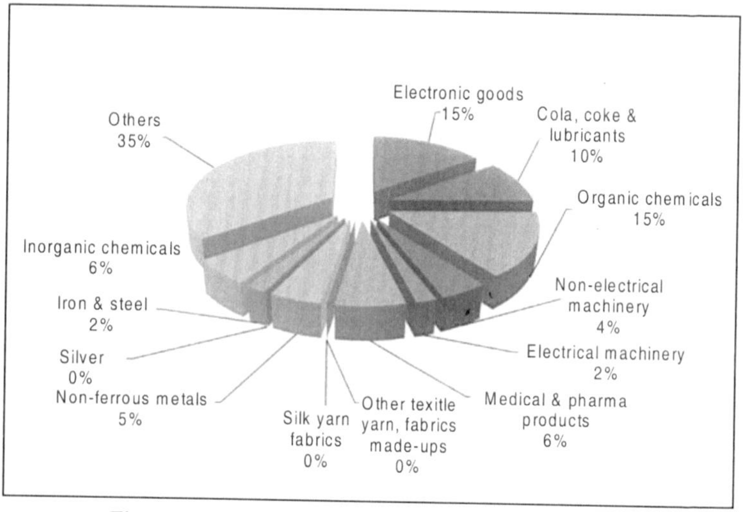

Figure 15.6 : India's Import of Principal Commodities from China (%) (1998-99)

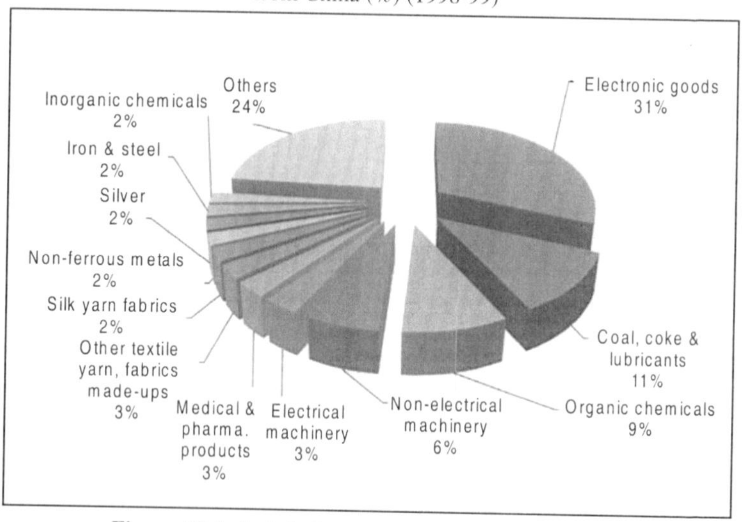

Figure 15.7: India's Import of Principal Commodities from China (%) (1998-99)

It is revealed from the above discussion that there is enough scope for increasing bilateral trade between India and China, which can be achieved by restructuring policy options in vital areas of co-operation. The long-term policy for commercial co-operation should be based upon the existence of support and complementarities in the economies of India and China to make them mutually gaining partners in trade and trade related activities.

REFERENCES

Batra, Amita and Khan, Zeba (August 2005): *Revealed Comparative Advantage: An Analysis for India and China,* Working Paper No. 168, Indian Council for Research on International Economic Relations (ICRIER).

Bhat, T.P., Guha, Atulan and Paul, Mahua (2006): *India and China in WTO-Building Complementarities and Competitiveness in External Sector*, Report of Planning Commission, Government of India, ISID, New Delhi.

Bhattacharya, Swapan K. and Bhattacharya, Biswa N. (April 2007): *Gains and Losses of India-China Trade Cooperation—A Gravity Model Impact Analysis*, CES Working Paper No. 1970, Category 7: Trade Policy.

Bhattacharya, Swapan K. and Bhattacharya, Biswa N. (2006): *Free Trade Agreement between China and India: likely Impact and its Implication to Asian Economic Community*, ADB Institute, Discussion Paper No. 59.

Bussiere, Mathieu and Mehl, Arnaud (January 2008): *China's and India's Role in Global Trade and Finance – Twin Titans for the New Millennium*, Occasional Paper Series No. 80, European Central Bank.

Debroy, Bibek (1991): *China's External Trade Since the Reforms*, Commonwealth Publications, New Delhi.

Dimaranan, Betina, Ianchovichina, Elena and Martin, Will (August 2007): *China, India and the Future of the World Economy: Fierce Competition or Shared Growth?* Policy Research Working Paper – 4304, The World Bank Development Research Group Trade Team.

Dzever, Sam and Jaussad, Jacques (1999): *China and India: Economic Performance and Business Strategies of Firms in Mid-1990s.*

Panagariya, Arvind. *India and China: Trade and Foreign Investment,* Columbia University, New York.

Report of the India-China Joint Study Group on *Comprehensive Trade and Economic Cooperation* (2005), New Delhi, India.

Sidhu, W.P. Singh & Yuan, Jing-dong (2005) *India and China — Cooperation or Conflict?* India Research Press, New Delhi, India.

Weerahewa, Jeevika and Meilke, Karl (2007): *Indo-China Trade Relationship: Implications for South Asian Economies*. Paper presentation at the IATRC meeting in Beijing, China.

16

Good Governance in the Panchayati Raj Institutions

After the 73rd Amendment of the Constitution of India and inclusion of Chapter IX and Chapter IXA in the Constitution, Panchayati Raj Institutions (PRIs) have become indispensable part of the governing system in India. There are two other countries in the World, *i.e.* Brazil and Nigeria who have provided independent constitutional status to their local bodies. The Panchayati Raj Institutions provide distinct advantages:

(a) The Panchayati Raj Institutions have in-built community participation, and
(b) There is not much expenditure involved in using this machinery for governance.

In few states, where the PRIs have long and stable existence like in West Bengal, Kerala and Gujarat, they have become part of the main governing system and behave in no different way than their counterpart in the bureaucratic set-up of the Government. Sometimes, they behave even in a more bureaucratic way causing avoidable delay and indecision.

Let us analyse how can Panchayati Raj System be made as a real support system to the governing mechanism in India. For that purpose, let us take following seven parameters to evaluate the governance by the PRIs in India:

(i) Simplified transactions or procedures or processes;
(ii) Autonomy of the organization;
(iii) Faster community participation;

(iv) Promoting competition;
(v) Effectiveness;
(vi) Mechanism for fixation of accountability; and
(vii) Political and administrative will.

(i) Simplified Transactions

In PRIs, the transaction can be concluded in a simple manner. A villager as a client or customer goes to Panchayat, talks in his local language which Panchayat authorities understand well and gets a solution or rejection then and there. The office bearers of the concerned Panchayat take no time as they are well aware of his status or eligibility and their own resources in hand and other competing persons. The panchayats need not maintain detailed record office like a bureaucratic office as their accountability is by the people and of the people who will be interacting with them for their lifetime. It is not like impersonal and indifferent way of working in other Government offices. Some Panchayat bodies have evolved their own procedures or procedures which are much more simplified and accessible to the people of the area. There is no administrative hierarchy or manual by which they are governed except in cases of embezzlement or misuse of public funds. They unfortunately have been held as 'public servant' under the Indian Penal Code and this has been upheld by the Supreme Court in few cases. Therefore, beyond a limit, they cannot go for simplification to the extent of not maintaining the insignia of records of expenditure. If one goes to a Panchayat office, one will not find lot of paper work except some registers maintaining lists of BPL (Persons Below Poverty Line), dispensaries, schools, villages, deep tube-well and other sources of irrigation, population statistics and other development indices and the shelf or bunch of schemes. Whenever they receive indication of scheme or fund, they send the readymade list after discussion in Panchayat body or in its sub-committee in a democratic way to higher Panchayat body or official hierarchy. After receipt of fund, they send the utilization reports, follow-up the beneficiaries and close that scheme. Thus, they are not to maintain complicated accounts or papers as they more or less function as implementing agencies.

(ii) Autonomy

To work efficiently, each organization should have requisite degree of autonomy. Now that the Supreme Law of the land has given them the constitutional status, they have adequate degree of autonomy in their constitution, composition, tenure, reservation of seats, taxes and funds, accounts and audit, jurisdiction and their powers and responsibilities. This is the highest status, which any institution in the country can aspire for. This has also given meaning to the spirit of Article 40 of the Constitution. Now, the states cannot afford to make PRIs weak and ineffective like in the past owing to reasons such as the failure to hold regular and periodical elections, representation of Scheduled Castes (SCs)/Scheduled Tribes (STs) and the women, lack of financial resources and inadequate devolution of powers and responsibilities. In fact, the Constitution has thus, made a three-tier Indian federation incorporating Eleventh Schedule in the Constitution, which lists wide variety of subjects including important areas to be handled by the Panchayats. The provision of a State Election Commission ensures that unlimited supercessions are kept at bay and the appointment of a State Finance Commission ensures adequate finances to PRIs. But the constitutional mandate cannot be fulfilled unless the State Budgets for the schemes with regard to the subjects listed are fully transferred to the Panchayats at the appropriate level and the adequate manpower and staff are placed under the control of panchayats. There should be provision of proper training of fresh Panchayat representatives after each election so that they are empowered to discharge their functions effectively. Autonomy without capacity building is meaningless. The powers and functions of each tier of Panchayati Raj bodies should be defined by the State Law and rules in this regard. More scholars, retired Government servants, lawyers and technical personnel should be associated in Panchayats so that they become efficient units of public governance and vehicles for development in their jurisdiction.

(iii) Community Participation

'Top-down' approach has failed in our polity and 'small is beautiful' is proving successful. Panchayats are by the community, of the community and for the community, they represent. The existence of the Panchayati Raj is in fact, an extension of democracy up to the village, and has found favour amongst politicians except with MPs and MLAs. Now participation of the people is no longer an ideological commitment, it is given shape through legislative and structured measures in creation and running the Panchayats. "For the last few years, there has been a sharp shift in the perception of the rural people towards the change in developmental activities. Political and social awareness have also opened the eyes of so far deprived sections of the rural society. Any talk of power to the people becomes meaningless unless power is exercised by the people at the grassroots level for the society and political awareness is channelised in the right direction. The constitutional provisions in Chapter IX aim at reducing political and bureaucratic interference in rural development programmes. For this, many States have redefined and redesigned the role of state bureaucracy in order to facilitate effective functioning of PRIs. In West Bengal, even District Collector is brought under the Chairman of the Zilla Parishad and Block Development Officer under the President of the Panchayat Samiti. Most of the departmental officers should be asked to be accountable to the respective tier of Panchayat body so that their performance can be factually appraised. At the same time, after the election, Panchayat bodies should not behave like Council of Ministers who becomes responsible to people only after five years or before, if dissolved earlier. They should function as units of decentralized and democratic administration and should adapt participative decision-making through sub-committee and through proper deliberations in them. Even Gram Panchayats have been asked to directly call meeting of all their voters/inhabitants at least twice in a year and should place their plan Budget, schemes and performance reports before them for approval. It should not be a mere formality and participation of villagers should be ensured by presence of third party during such meetings.

Panchayat Samitis and Zilla Parishads should also publicize their plans, programmes and performance reports on notice boards, through local newspapers and even through District Publicity Office. There should be complete transparency in their functioning. The 'Pradhan' of Gram Panchayat should not move with the savings pass book in his pocket just because he has been so authorized by the State Government to spend the money. He should take the opinion and suggestion of the villagers into account and implement the schemes incorporating or rejecting their suggestions after due consideration and discussions. The model of community participation should be implemented at each tier of Panchayat as far as possible. Even the villagers should be allowed to share usufruct or profit of any community based schemes or projects, which have been implemented without the funds placed by the Central or State Government. It is only Panchayat which can promote self-supporting schemes by the villagers through their '*shramdan*' like water harvesting schemes or social forestry, etc.

(iv) Promoting Competition

Panchayats cannot afford to be isolated from the present socio-economic environment of liberal and market economy. Accordingly, the competition is essential to ensure efficiency and to avoid subsidies. The Panchayats are not all pervading and are not sovereign. They have to compete with the bureaucratic regime in the same field. Therefore, it is advisable that Panchayats and other departmental agencies are allowed to operate in similar field to a limited extent. For example, under PMGSY (Prime Minister Gramin Sarak Yojana) and NREGP, the Public Works Department (Roads) may be simultaneously allocated some schemes so that people can compare implementation of schemes by two agencies. This itself will be an eye-opener for both the important governmental agencies. In a similar vein, the Panchayats may be allowed to run some schools under their umbrella *vis-à-vis* schools run by Department of School Education. Panchayats may not try to extend their arms to regulatory and unexplored fields where they lack experience and expertise. They should concentrate on rural development schemes specially schemes which are labour intensive and influence the lives of the villagers directly like employment generating schemes such as construction of rural roads, pisciculture, small irrigation, agriculture, education and health.

The Panchayats in West Bengal and Kerala are successfully associated in land reforms, literacy drive programme and village sanitation programmes.

(v) Effectiveness

Panchayats should be confined to their micro-level development plans. They should concentrate on mobilization of resources through local resources like taxes and fees and by generating profits in the assets or schemes promoted, constructed or maintained by them. If they try to spread their wings like operation of hospitals and schools, they will lose their main focus on rural employment programmes or rural assets creation schemes. They can look for help and co-operation from NGOs and even from donors in India and abroad. They should not waste time in criticizing other States or Central Government. Rather, they should try to become self-dependent or self-supporting. If technical expertise is not made available, they should try to avail from the open market. The system of Panchayats has been enshrined in the Constitution to provide them little more freedom from the rigid bureaucratic rules, regulations and cumbersome procedures or processes. They are accountable to their electoral college and none else. Their aim, therefore, should be to deliver gains of rural development programmes to the electorates irrespective of the caste, creed or political colour of the beneficiary. A poor is requiring food, shelter and clothes irrespective of the party he belongs. After being elected, a Panchayat functionary is a representative of the whole population and not only of those who voted for him. If this mission is adopted, effectiveness will automatically be generated. If he targets his voters or supporters only, he could never be effective because he has already avoided a sizeable percentage of beneficiaries from the beginning of the scheme or programme. The panchayats in West Bengal and Kerala were successfully associated in land reforms, literacy drive programme and village sanitation programme. The Panchayats in West Bengal even control communal flare-ups and commission of crimes in villages by mobilization the community against such acts of violence. Sometimes, the response becomes more violent than the crime committed. Such kind of self-acquired job by Panchayats should be discarded. They should concentrate on 30 per cent persons below poverty line, weaker sections of SCs, STs, OBCs and women. The women representatives which constitute 33 per cent including 33 per cent Panchayat officials, are enough to safeguard against any injustice to women and can ensure them equality and safety in the community. In this age of competition, an ineffective organization will die its own death and will be soon replaced by another organization. Bureaucracy is awaiting as their natural substitute. The Panchayats can effectively promote the overall development of rural areas as follows :

(a) by implementing social development programmes,
(b) by reconciling the priorities of decentralized planning with those of national and state planning,
(c) by playing a bigger role in implementing self and wage employment generating programmes which have the potential to promote community and social development, and

(d) by taking initiatives which would ensure greater equality in development.

(vi) Accountability

The democratic organization or process loses all its credibility when there is no accountability. The responsibility for utilizing the tax resources and fund placed by the Central and State Governments to the maximum extent possible would devolve on the PRIs in the States.

The accountability is in-built in the Panchayat organization. At Gram Panchayat level, the neighbour and elector of a Panchayat member would be the first person to know what facility his elected representative is enjoying whereas a Government officer is not accountable to anybody who runs away with his family and ill-gotten fortunes when he is transferred from the place of posting. Though the accountability of Panchayat Samiti and Zilla Parishad representatives is more far-fetched, yet it is always more concrete and is closely monitored by their clients than in the bureaucratic organization. They are at least accountable to their beneficiaries or clients after every five years whereas a Government official though accountable on paper, cannot be made responsible for his misdeeds or misuse of authority in the present system of bureaucracy. The accountability should be with respect of both, *i.e.* performance and financial matters. The XI (Eleventh) Finance Commission has clearly laid down that the requirement of financial accountability of PRIs should be designed and supervised by the Comptroller and Auditor General (CAG). The accountability of Panchayat units should not only be with the next higher level or levels of panchayat but can also be towards an independent bureaucratic authority. Each person or authority who spends or utilizes single paise from the Government exchequer is a public servant and is accountable to the people who have contributed their hard earned money in trust with the Government or Panchayat in the form of taxes or fees. The systems of internal audit and third party audit by the Accountant General have been laid to affirm trust of people. Therefore, the Panchayats should also be accountable to them apart from their superior Panchayat bodies. Few states have provided mechanism of accountability of Panchayat bodies of three tiers. For example, in West Bengal, Sub-Divisional Officer or SDM can inspect the Gram Panchayat, take action against any Panchayat member, Pradhan or Gram Panchayat as a whole in the form of expelling him as Panchayat member or termination of his membership or filing the cases against them in the police or in appropriate court for defalcation of fund. Similarly, District Magistrate should be empowered to inspect the functioning at the Panchayat Samiti or Mandal Panchayat level like it has been provided in Karnataka.

On the pattern of Maharashtra, Divisional Commissioner should be having power of control of Zilla Parishad including inquiry into their affairs and expulsion of member or President or the Sabhadhipati if he indulges into misfeasance, *i.e.* misuse or misappropriation of Zilla Parishad fund for his personal illegal gain. The State Vigilance Commission should be authorized to go into complaints of corruption against Panchayat bodies or their members or office bearers. It will not only ensure better performance by panchayat bodies but will also inculcate transparency in them. Government will then not have to create another body for the supervision and control of Panchayat bodies separately and people's problems will be solved locally. A villager is not supposed to complain at the state level. However, Government officers should not question the decision-making process of decision itself. They should be allowed to question the offences of corruption or misutilization of fund or misappropriation of fund only. It does not include non-submission of utilization certificates or reports. It does not mean that *bona fide* acts of Panchayat bodies will fall under this category for the sake of accountability. They have been empowered by the Constitution to manage affairs of the people locally. The fixation of accountability should aim to increase their efficacy and transparency rather than whittle down their working or choke or obstruct them in their due performance of duties in the broader interest of the local public, they serve.

(vii) Political Will and Administrative Will

There are certain prerequisites for the survival of the PRIs. They are :

- There should be commitment to the Panchayati Raj as a political value and the consequent political will to establish it and to sustain it in proper discharge of their functions;
- The effect to revitalize Panchayati Raj, in fact, is an attempt to construct the Indian polity by way of bringing so far deprived sections into the mainstream; and

- The Panchayat bodies should be treated as the single agency for developmental work at the rural local levels to satisfy people's needs and shall meet the hopes and aspirations of the people, they are serving.

The Panchayat system is a supporting leg of the administration of the State. Panchayat bodies are offsprings of state political executive. Therefore, they cannot function without positive and willing support and patronage from the State political executive reflected in the Council of Ministers of the ruling political party. Similarly, the Panchayat bodies should enjoy unequivocal support from the political apparatus of the ruling political party otherwise they will be continuously disturbed by the Ministers or Members of State Legislatures because they are eyesores in perpetuation of their hegemony in respective areas of MLAs. After the Panchayat bodies are created, most of the needs of the local people are met by the Panchayat and people stop looking for MPs or MLAs and their power of patronage is eroded. Similarly, bureaucratic administration becomes hostile because they lose their reference power as administrators within their jurisdiction. Thus, only a stable political will can help the Panchayat bodies to function properly and deliver good to the people.

Similarly, administrative will should permeate local Panchayats and in administrative hierarchies to see that these local administrative units created by the Constitution function properly. The Panchayat bodies should see that they are open minded, take suggestions from all sides and implement the programmes in the best interests of the local populace. The Panchayat Samiti and Zilla Parishad should be very liberal to provide technical assistance through their technical personnel at hand and through sophisticated equipments at their disposal to the Gram Panchayat and the Panchayat Samiti respectively. Similarly, the engineering departments like Department of Public Works, Irrigation, Minor Irrigation and Public Health Engineering should come out on their own to provide technical guidance and the field support to the Panchayat bodies so that they cannot only utilize the funds allotted to them but may also create sustainable assets out of the rural development fund released to the PRIs. The Departments should not indulge in line hierarchy but should expand it horizontally so that Panchayat bodies may use their expertise at the grassroots level. The Panchayat bodies should use the Government staff and officers in the best interest of the implementation of schemes rather than just Lord over their local level Government functionaries. Then they will be no different than the feudal Lords in the bureaucratic hierarchy. This will then be another line of bureaucratic set-up and will lose its edge to deliver instant solution to people, they are meant to serve. However, without effective control of the Panchayat representatives by their political parties. However, the system will go into disuse soon.

Achievements

Panchayats so far, have performed well. Due to entrenched Panchayat system, Left Front in West Bengal has been coming in power for about last 30 years. The women representatives have come in the forefront and gender gap is being obliterated automatically without adopting any other programme for that purpose. There are 2.5 lakh elected representatives of Panchayati Raj Institutions along with the urban local bodies who constitute an army of some 32 lakh elected representatives, with $33^1/_3$ per cent women representatives and proportional representation of Scheduled Castes and Scheduled Tribes and Other Backward Classes. 'It is political and social improvement on a scale without parallel in the World and without precedent in history', according to Union Minister of Panchayati Raj, M. S. Aiyar, 23-25 July 2006, New Delhi. The people's satisfaction has come to the light through various publications, press and media. The very association of the rural people has given them satisfaction of living in a real democracy where they are makers of their own fate and future. The deserving rural poor have got benefits and the schemes which have been implemented by PRIs suited more to them than the previous ones implemented by the bureaucratic regime. Based on local experience and expertise, sometimes a little expenditure has done wonders. Panchayats have taken many local schemes from their own funds which could never be done before because there were standard schemes without any provision to customize. Schemes specific to that locality can now be executed freely. Rural folks have been inspired. They have executed check dams or water harvesting schemes with their own contribution of labour which have changed the environment of entire village. A drought prone village has now become green and the rural population does not need to migrate to other distant places resulting into their exploitation.

Most of the villagers now have their houses built by Panchayats under various rural development schemes. They avail employment generation schemes and incidents of starvations have almost been eliminated. Times are not far when they will be isolated from the clutches of the local landlords and moneylenders. The oppression of weaker sections and *dalits* has been reduced. The rural folks have realized their power and potential. Time is not far when they will not be burden but will lead the Nation in the march for prosperity in the 21st century. To tackle this problem, retired people who have appropriate experience in the relevant field can enter into Panchayat body politic as full timers. Their valuable services can be utilized at no cost. However, young and bright men are always preferred if they can afford time to manage affairs of Panchayats during holidays or otherwise without causing prejudice to their present jobs.

Lack of technical inputs is another drawback in the functioning of panchayat bodies. Technical inputs must be provided to the Gram Panchayat by the Panchayat Samiti and to the Panchayat Samiti by the Zilla Parishad. The concerned department of the government should also provide latest technology available through their respective hierarchy specially agriculture and engineering departments. The panchayats should harmoniously synthesize their local experience with the latest technological innovations in the form of scientific equipments or instruments of works and tools of information technology to provide all information to villagers at their doorstep and to communicate or send reports at the higher level of hierarchy subject to the condition that the technological innovations are assimilated by the people for whom they are meant. For this purpose, extensive training and extension work at each tier of Panchayat is equally important.

Time consuming decision-making is another negative factor in the working of Panchayats. All decisions in Panchayats are taken in a participatory manner through meetings. For this purpose, notices are to be given, meetings are to be arranged, minutes are to be drawn up and signed and are to be communicated for implementation. This takes lot of time. This time can be reduced if action is initiated like tendering etc. or procurement process in meeting. Every Panchayat body should have shelf of schemes ready in his bank of schemes so that whenever funds are received or green signal for funds is received, they can go ahead with implementation of that scheme. In case of non-receipt of fund, such scheme can be booked under their own fund. Funds take time to travel up to the grassroots level. However,

Central Government has been giving some funds directly to the Gram Panchayats but it is not enough. Cutting or bypassing the stages or channel is always a good practice provided utilization reports are ensured. This is also another field where the Panchayats have to develop their in-house expertise.

Thus, delay is the first enemy of the Panchayat bodies, corruption the second, partisan functioning the third and lack of technical skill is the fourth enemy in order. I am sure that Panchayat bodies will remove the hindrances of their enemies in their path of progress and will work hand-in-hand with the administration to ensure that India gets its rightful place in the comity of four/five top Nations of the World.

During the interactive meeting with the Clean Village Award winners/Chairmen of Panchayats, former President of India, A.P.J. Abdul Kalam had administered the following oath to them on 24 March, 2006 :

- I will work for the development of my village and upliftment of the people irrespective of any affiliation.
- I will assist at least two neighbouring Panchayats in establishing an integrated sanitation mission in their villages. I will ensure that the people of my village do not spend their hard earned income in alcohol and gambling and also that the youth do not become the victims of narcotic and psychotropic substances.
- I will work for making my Panchayat totally litigation-free and ensure that all our differences are settled amicably within the Panchayat itself.
- I will work for making my Panchayat 100 per cent literate.
- I will work with the members of my Panchayat and plant at least 10,000 trees in our area during the next three years through people's movement.
- I will ensure that at least one member of each family in the Panchayat will be trained, empowered and provided productive employment using the core competence of the village and the assistance of district authorities.
- I will be transparent and righteous in all financial transactions connected with village development.
- I will be a role model through my way of life.

I am sure that representatives of Panchayats will follow the advice of the President of India and make Panchayats an ideal machinery of rural development in India.

Conclusion

Panchayats are now a reality in India. Now, the issue is how to make them efficient and effective so that they can come up to the expectations of the people. It is a fact that through the system of Panchayats, a parallel implementation machinery has been in place in the country without incurring billions of non-plan expenditure on them which is a major problem in the development of the economy in the country. The Panchayat functionaries or office-bearers work without remuneration. Even for meetings, they get negligible remuneration. Therefore, some corruption has crept into the arena of Panchayat bodies. However, those Panchayat functionaries are exceptions who work for ulterior purposes. These people are in every field. The corruption has to be abhorred at all costs. There should be zero tolerance to corruption of all forms to make India stand in the present competitive era of World economy propelled by information technology. Corruption is just like an infection which, if started, does not stop at one destination and goes on to newer and farther destinations.

REFERENCES

Aziz, Abdul : *Decentralized Governance in Asian Countries*, Sage Publications, India Pvt. Ltd., New Delhi.

Arnold, David D. (ed.): *Market-I*, New Delhi-110048, year 1996, Total page-301.

Gupta, M.C. and Tiwari, R.K.(ed.) (1998): *Restructuring Government*, Indian Institute of Public Administration, New Delhi.

Jain, S.P. and Hochgesang Thomes W. (ed.) (1995): *Emerging Trends in Panchayati Raj (Rural Local Self-Government) in India*, National Institute of Rural Development, Hyderabad.

Naidu, N. Chandrababu and Seventi Nivan (2000): *Plain Speaking*, Viking by Penguin Books India (P.) Ltd., New Delhi.

Panchayati Raj Update (Vol.-July 2006): Ed. Rajkishore for Institute of Social Sciences, New Delhi.

Reddy, P.L. Sanjeev : *Issues and Themes in Indian Administration*, Indian Institute of Public Administration, New Delhi.

Singh, J.L.: *50 years of Panchayati Raj & Rural Development*, Manak Publications Pvt. Ltd.

Singh, S.S., Suresh Mishra and Sanjay Pratap (1997) : *Legislative Status of Panchayati Raj in India*, Indian Institute of Public Administration, New Delhi.

Wadhwani, M. and Mishra, S.N. (ed.) (1996) : *Dreams and Realities : Expectations from Panchayati Raj*, Indian Institute of Public Administration, New Delhi.

17

Leadership in Globalization

The leadership in the liberalized economy has to be objective, fast moving, knowledgeable and transparent.

Leadership has been defined in various ways. It has been proved beyond time and space that each one of us can be leader in our own position if we act with motivation. The leadership is a fire within which comes out as desire and then converts itself into undying aspiration. Then it goes through execution process with drive, dedication and undaunting spirit. It surpasses all hardships and obstacles. But leadership is not only emotion. It is understanding and knowledge with definite purpose. The purpose or aim is broad and encompasses not only the well-being of the leader but the welfare of the community as a whole. "Leadership is a term used to describe the act of transforming, inspiring, mentoring, coordinating and managing people towards individual's, group's, an organization's, a community's a nation-states vision, goals and objectives." Leader is a man or woman who influences the beliefs and behaviour of a group. Leadership is one of many types, *viz.* autocrat or tyrant, reactive, participative. Weberian model based on rules and regulation like bureaucracy value-based, personality cult-based, charismatic or informal.

Let us attempt to define the amorphous concept of leadership in a concrete way with the help of six factors or components following mathematical formula below:

L/100 = Vision (V/15%), + Self Example (E/25%) + Knowledge and Authority (K/15%)+ Participation and Communication (P/15%) +Work (W/20%) + Energy E/10%)

In addition to above addition formula, we can attempt to approach leadership through simple multiplication formula known as Cobb Doglus formula as:

$$L = \text{square root of } V^3 \times E^5 \times K^3 \times P^3 \times W^4 \times E^2$$

Thus, the leadership is multiple function of above 6 (six) factors. The addition formula is simpler and easier and conforms to most of the field verifications. Therefore, let us further firm up the first formula by introducing a variable in each factor so that each factor can be concretized.

Thus, as per factor analysis technique, the linear equation can be thus formulated :

$$LS_i = VS_1 + ES_2 + KS_3 + PS_4 + WS_5 + ES_6$$

The variable or co-efficient S_i may be determined for each factor depending on the relative importance of that factor. Here we cannot avoid subjectivity. For this purpose, however, a rating scale may be fixed. The rating scale, by and large, will secure the intensity of the assessment on a pre-defined sets of response alternatives. The rating is fixed on the basis of Likert scale which is also called a 'summated ratings' scale. Let us assign the numerical value or rank ranging from 10 to 0 to different intensity as follows:

Very good	assigned	10 marks
good	assigned	8 marks
average	assigned	5 marks
below average	assigned	2 marks
nil	assigned	0 marks

Using this method of scoring, total scoring is calculated depending on the strength of the factor.

Let us therefore, assign percentage importance to each factor taking clues from the following discussions.

(a) Vision

Fifteen per cent weightage is given to vision of the leader. The leadership should contain visionary ideas which other members of the organization cannot even comprehend. Each organization functions on a mission. The mission has to be very broad. It should involve in ultimate analysis, the welfare of the society or the nation or the World. After the mission is fixed, the objective or aim of the leadership is prescribed as subset of the mission. The vision or mission is defined for a long period. After that aim is defined, then the targets are fixed for five years or annually. Therefore, the organizations which are only profit-motivated can only be categorized as average with strength factor 5. This is applicable in most of the private enterprises or firms. However, where corporate goals are fixed like for improvement in environment, fillip to the economy and redistribution of profits and wealth of the organization among shareholders or employees, improvement in infrastructural facilities, sustenance to the health of the community, supplying drinking water to the community can be categorized as 'good' with a strength factor of 8. In case of Democratic or Socialist Government where the goal is to completely disseminate the Government revenue for the all-round development of the public, the strength to the factor can be assigned as 10 provided the delivery mechanism is also suitably placed in tune with the objective.

In another category, those regimes will fall which are controlled by the dictator, or Sheikh, military junta, familial king or ruler who have complete hold on the public through their brutal force of the state and have throttled the voice of the public. In such organizations, the freedom of speech and expression is generally not there. Even in democratic societies, some leaders are visionary whereas others occupy the power by hook or crook or through networking or unhealthy means. Such organizations will be given only 'average' weightage. Firms or proprietorship organizations, where there is sole objective to make profits, will come under category 'below average'. But where the individual firms or organizations and their masters have goal to run their livelihood with modest profit-keeping in mind the interest of the customer, or clients moderately, they will jump from 'above average' to the category 'good'. A leader cannot afford unless his organization sustains. Therefore, they are allowed to ensure their sustenance first as their basic right and then they can part with fruits or pass profit, or facilities to their customers or consumers from time to time depending on the macro social and economic and business atmosphere. The leaders like Sri M.K. Gandhi, Pandit Jawaharlal Nehru, John F. Kennedy, W. Churchill, Jamsedji Tata will be given full marks. The dacoits, illegal gangs, mafioso or dons or terrorist leaders though running their organizations with efficiency, will be given 'zero (0)' mark because their goal itself is illegal. However, there is always a silver lining in their goals. They will then get some weightage for their way of working and for distributing their booty or spoils among the poor and the needy though they fail miserably when they are assessed in respect of other factors. A goal or mission of the organization cannot be allowed to be unlawful or illegal.

(b) Self-Example

In the Indian set-up, this is considered to be the most important factor in constituting leadership. A leader has to be first motivated himself before motivating others. This is a universal principle equally applicable during all times. A leadership cannot be exercised simply by giving sermons, messages or theoretical lectures. It will be an imposed kind of leadership which will wither away soon. In the age of emergence of people's voice through easily accessible media like T.V., newspapers etc., the public, workers or followers claim right to follow the example of their leader. A leader cannot preach integrity in the organization if he himself or through his cronies is dishonest. Similarly, a leader cannot ask for hard work if he himself indulges in gossiping during office hours. A commander cannot effectively lead his men to die unless he himself leads his contingent or formation from the front. This is also true in case of technical knowledge of the leader. In case of highly technical or skilled organization like Information Technology Multinational, the leadership also flows from technical foresight, planning and problem solving skill of the leader of the group or company. A sage or teacher cannot ask his disciple not to eat sugar, when he himself takes sugar. Gandhiji had led his countryman in India in a lone cloth because at that time the poor man had hardly a cloth on his body. Therefore, leader should not only be honest, sincere or skilled but he should also appear to be honest, sincere and skilled to the members of his team or clients or subjects he is dealing with. The Indian Army is famous for its bravery throughout the World because the commanders from top lead the bottom ranks of their men or *jawans* from the front unlike in other armies.

As a result, in any operation, casualties of army officers is more than the officers killed in action in case of other armies. Similarly, the Indian software engineers and managers are successful all over the World because they do not care for their personal comforts as compared to engineers or managers from other countries. As a result, they get readymade support from people working with or under them and willing co-operation from their clients. The Russian Cooperative

Farming System failed because the managers or administrators started sitting in isolated air-conditioned chambers while the workers worked in the field.

(c) Knowledge and Authority

The leadership without knowledge and authority cannot sustain long. In fact, a leader takes up a field in which he is an expert or has propensity to enter. It is equally important in technical field. Though large commercial or manufacturing companies have Chief Executive Officers who are not from the same field, yet he is definitely expert in other fields like human resource management or finance. These two fields are equally important for any corporation. Corporation is a living identity and requires support from all sides like a human being. Therefore, a leader cannot be confined to a narrow field. Sometimes, the national leaders have come from the field of science and technology and then have done far better than the one, who were from law or finance streams.

If a leader has authority over the people or workers, he can exercise the proper control or supervision over the organization. This authority should emanate from the legally constituted forum like body of shareholders in a corporation or company and free and fair election in a country or society. The authority becomes, more appropriate if it is exercised after proper information with due knowledge of details.

Thus leadership is assigned 15 per cent weightage for knowledge and authority. Sometimes, a leader may lose his hold over his organization but may continue with the help of a court order. This is not the authority. The authority should mean the willing allegiance by its constituents or units and not forceful or despotic control. If a leader has sound technical knowledge and authority, he will be categorized as 'very good' with 10 making. If the leadership is a joint decision-making body, it will be termed always in 'good' category because this decision-making process is dilatory and therefore cannot be termed 'very good'. Secondly, a joint body cannot exercise full authority but the decentralized authority cannot fail easily also. In this way, this type of decision making or power sharing will always be in 'good' category. If it has speedy decision-making, it will achieve 'very good' category.

If the head has assumed the charge just as a successor due to his family background or organizational rule, he will be kept in 'average' category to begin with. If after assuming power, he demonstrates, his calibre, and acumen, proper knowledge and skill, he will go up in ranking. If he starts administering or managing through his limited cronies or 'kitchen body' or cabinet, he will be 'below average'. If the leadership is intoxicated in power by his flatterers around and pays no attention to details of management and doesn't exercise his personal authority, it will get 'nil' ranking. His delegated subordinates may exercise his colourable power for their selfish ends and ultimately his authority will be eroded and eclipsed like the East India Company. There is no trace of the E.I. Company at present in U.K. Such types of instances are noticed in companies or regimes which become unwieldy or very powerful as much as power goes beyond their heads as absolute power corrupts absolutely. Therefore, there is no room for a leader to be complacent and to sit on his or her laurels and relax. He has always to go miles. He has to go for variations. The thing is 'good' today may be termed 'average' or 'bad' after lapse of time because the change is continuously happening. A leader has to move with time and look into the future.

(d) Participation and Communication

Leadership should involve every member of his team to achieve the goal. This is possible through positive and effective communication system in place. The leader should take the team along with him. It is possible that all members of his team or organization may not pull with him and are drag on the organization. But the true leadership is to develop the members of the team so that they can give their hundred per cent to achieve the goals set in the organization. Continuous training and retraining is one way. Second is constant monitoring and review of the performance of the employees. However, in case of appraisal of employees, the leader has to be objective and should be impartial to everybody. The leadership goes out of its way to make sure that employees enjoy their work. There should be mechanism in the organization to reward good work done or any contribution made by the employees. Employees are encouraged to experiment with new methods and try out creative ideas. The behaviour feedback is given to employees which they take it seriously and use it for development. Employees are told their strengths and weaknesses from supervising officers and colleagues. Employees returning from training programmes are given opportunities to try out what they have learnt.

A leader sits periodically with his team and communicates strengths and weaknesses of the organization. He puts forth the balance sheet before all the members even before his clients, customers and other stakeholders. He regularly communicates with his organization through a house magazine, journal or website and invites their feedback. His transparency helps to get willing co-operation and utmost devotion from all members of the organization.

A leader will be granted cent per cent if he is fully transparent in his work. This generally happens in Government organizations and professionals run companies or corporations. However, in case of smaller organization or family run companies, CMD or proprietor has the tendency to conceal each and every details, or transactions specially financial ones, so as to siphon off the profits for his personal benefit or for the benefit of his family or group. Thus, these companies or groups will be put in 'good' category. If they further manipulate, they will be in 'average'category. The companies run stealthily by the MD individually will be in the bracket of 'below average'. However, in case of no disclosure of annual report or return on financial transactions and other transactions, the leadership will get 'zero' and accordingly. They should not expect to get willing co-operation from employees. Similar is the case in respect of Human Resource Development, if the HR climate is personalized, the organization will be weighed only half for disclosure of financial transactions. Thus, the weightage factor is equally to create objective and development oriented climate and transparency in financial and other transactions of the company by the top rung of the leadership which will percolate down and will be applicable in a similar fashion up to the bottom layer of the leadership in a slightly changed manner.

(e) Work

Leaders convert motive into action. Action is thinking plus doing. Therefore, work is the means to achieve end. Leaders are achievers. If the leaders cannot deliver output, he would be overthrown by their team. A man can't remain static. He is subject to constant changes occurring around. Leader uses this universal principle to transform his organization and engage his team by way of diversification, rotation or by empowering them so that infinite energy is released out of work to achieve the goal. And his organization survives in spite of changes in future. He knows the art of doing. He hates inaction. A dull person engaged in work is better than an idle intellectual person for the society. A leader must make common man do uncommon things. If timely action is not initiated in any field, the very purpose of doing will be lost.

The leader not only himself has exceptional capacity of work, he also demonstrates to his workers or team that he is in fact working constantly. But the real test of the leadership is that he can get the work done by the members of his team or the people, he is leading. He has to extract work form his subordinates by bringing out best of them. It is just possible that his people don't know what great potential they have to give to the organization. The leadership helps them to realize their full potential and channelizes for achieving the goal of the organization. Work or '*karma*' is very simple. According to the 'Bhagwad Gita', the work is conserved. A person will have result of his work one day or the other yet one should work without caring for fruits. It doesn't mean that one shouldn't plan for the work. One should very well plan his work with acumen so that delay in performance doesn't occur. There are three kinds of works—negative work, work, positive work. If the leader can ensure that negative work is reduced in the organization to the minimum and positive work is carried out to the maximum, he can be termed as a successful leader. He shouldn't disturb average worker who for whatsoever reasons, *i.e.* background, prejudices or lack of education or training, cannot change or update himself. They are the permanent base of an organization who are devoted for the well-being of the organization without much expectations. However, this does not mean that the leader should not experiment with new methods of work, new technologies, mechanization or reduction of cost. He should keep the organization constantly vibrating with the new technological developments in its field like use of information technology etc. in spite of opposition from rank and file. Technology can do wonders. It is God Viswakarma who can solve any problem. Therefore, training and retraining is one field which leadership should pay due attention.

However, there is a limitation for leadership. The leadership cannot wipe out the influence of the macro environment. For example, though India is a land of work (*karma*), the work culture in India is lagging behind China, Japan, USA, UK, France and Germany. The Japanese have high motivated workforce in the World because all work is done under the same roof. There is no discrimination between an officer, supervisor and employee or worker. The worker works for the organization and not for self. Idle colleagues are condemned. They have adopted the principle, of Kaizen, *i.e.* continuous improvement and TQM (Total Quality Management), following the Buddhist culture from India and adapting according to their environment. As a result, the Japanese workforce controls strains, avoids inconsistency and waste. Similarly, the Chinese following the Buddhist and the Marxist philosophy of Mao Tse-Tung, have surpassed every other country in the World because the workforce gives more output by taking least wages among the developed countries. The Socialist system allows every bud to blossom. The key of success of the Indian multinationals in the World like M/s Laxmi Mittal, Tatas, Infosys, TCS is that they employ top management and workers from India at a lesser compensation than their counterparts but they are more experts and give more and better output as compared to their colleagues from any other part of the World.

Here, it will be unfair if the contribution of the Indian Public Sector is not mentioned to make large trained workforce available in India who worked under large number of constraints of Government bureaucratic set-up. They have been properly hired by the Indian multinationals and utilized to spread their wings internationally. On the other hand, US multinationals like Enron, Lehmen Brothers, General Motors, City Corp have failed because of high salaries, perquisites and expenditure by their top executives which had the dampening effect on the motivation of their workforce. The Japanese and German managers have lived up to their reputation and have not been spoiled by bulging profits of their organizations. In China, the same is out of question because the wages of the leaders are fixed by the leaders of the country and one has the compulsion to accept the same without reservation. This may be subject to debate but China's progress in every field is undisputed. It is just possible due to devotion of work and better output that Chinese companies will rule over the World markets in times to come. Unless the corrections are made by the leaders and workers of other countries, they will be mute spectators to the China's leap to the World leadership. It is high time that work culture is infused not only in leaders but also in public. A worker, if criticized at his home or by his peer for hardwork will cease to work in long run. Therefore, a work climate is to be created which will inform all the sections and departments of an organization. Then, there will be quality output and conducive environment in each organization. The leader who is absorbed in work fully along with his followers or subordinates will be rated as 'very good' because he then is motivating his employees or people by his example.

The leader who works himself utmost but is not able to motivate his team fully will be graded as 'good' with 8 marks. The leader is one who himself performs and sees that others also perform.

The leader who is routinely doing work during office hours without making the work interesting for his organization, will be valued as 'average' whereas the leadership which is only engaged in planning and idle gossiping will be 'below average'.

A leader who just shows alibi for work, is not respected by his workforce and is ultimately rejected by his people and he faces his downfall and causes gradual but certain failure of the organization, he is heading. He will, therefore, be given 'zero' marking.

(f) Energy

The leaders discover energy to achieve their goal. Unless there is a fire or an urgent urge to achieve, leader cannot be born. According to Debashis Chatterjee, leader see their world within themselves. This is close to the Indian thinking which says that every being has infinite source of energy because God himself is seated within it. There is no difference between the God and men's inner master or controller called *atman*. The followers expect infinite energy in their leaders. Sometimes leaders grow old yet the followers expect God sent energy in them. They talk a lot of about their leaders how he works tirelessly for hours more than any worker. His time management is the talk of the organization including his leisure and hobbies. Therefore, leadership is a very cautious path which one has to trace, if one chooses to be the leader. At least his 50 per cent personal life is off. He has to sacrifice his privacy and personal whims and fancies for the sake of success of the organization, he is working with. His one bad example can have lot of repercussions on the people working in his organization.

One has to pursue one's goal stubbornly. One has not to retreat after failures. One has to try hard and should finish the work whatsoever has been started.

The Japanese people have attained the highest rate of economic growth in the World. This is due to their commitment to their daily work which they regard as sacred. Consider the following song which workers in the Matsushita factories sing before starting their work everyday:

"For the building of a new Japan,
Let's put our strength and mind together.
Doing our best to promote production,
Sending over goods to the people of the world,
Endlessly and continuously.
Like water gushing from a fountain,
Grow industry, grow, grow, grow!
Harmony and sincerity!"

Life becomes meaningful and interesting only when one loves to work which is constructive and challenging. JRD Tata worked up to his last day in life. Montague Norman worked as Governor of the Bank of England for twenty-four years from 1920 to 1944. He was completely absorbed in work. There was no room in his mind for worries and he could live a meaningful life. Bertrand Russel said, "The wiseman should wish to die while still at work". Therefore, now-a-days, the young leaders or CEOs are preferred over the old ones because they can plan for future. They have great stake in future. They are not feudal and are objective. They decentralize the work and delegate power empowering leaders at every level. They are open minded and promote informations and foresightedness in the organization. All great leaders thus work tirelessly for years together without being tired like Sri Jyoti Basu worked for 27 years as the Chief Minister of West Bengal. Pandit Jawarharlal Nehru worked for about 17 years as the Prime Minister of India. Similarly, George Washington and Abraham Lincoln worked for many years as the President of USA. The Prime Minister of any country, for instance, is the hardest working person facing unrelenting pressure of work day in and day out. A leader generates energy around. Don Burr of People Express Airline in 1980 created highly motivated workforce which could be rotated by way of participatory methods. Each leader at any stage of the ladder should be like the Prime Minister of his country because he also has twenty-four hours time at his disposal. All leaders have authority. Authority without energy to exercise is meaningless.

Thus the leader, who has vast energy, will be rated as 'very good'. The leader who complains or goes on leisure frequently can be put in 'good' category (mark 8). A leader who is working more than the general workforce will be placed in 'average' category with 5 marking. A leader who is working even less than his subordinates or followers will be in the 'below average' category. The leader who is known as idle by his group will get a big zero.

Concluding Remarks

The foregoing attributes enumerated are very strong ones. Anyone attribute of leadership in full bloom can make a leader successful. For example, Brahma, Jesus Christ, Prophet Mohammad, Buddha, Mahavira, Ramakrishna, Vyas, Tulsidasa, Karl Marx were leaders because they had vision. Similarly, Steve Jobs who founded 'Apple Computer Inc' had a vision to use computer as a route to changing the World. Similarly, when Welch became the Chairman of GE in 1981, he had vision to make GE as highly competitive company by massive downsizing and making it number one in the World. John Akers, CEO of IBM has given all emphasis on customer's satisfaction. Lee Iacocc, CEO Ford changed to Crysler Corporation to improve product quality and to recruit low-paid executives to cut costs to the company which made progress later on. Ramakrishna, Vivekananda, Gandhi, J.R.D. Tata, G.D. Birla, Ford, Bill Gates, Narayana Murthi and Laxmi Mittal were successful because they demonstrated the power of self-example. Gandhiji is the greatest example of self-example. He did not preach anything which he didn't practice. As a result people on their own followed Gandhiji and he dethroned the highest power of that time through non-violent movement. In the Indian Army, commanders die first than their men. The men eat first and officers eat last in the Indian Army. All the great leaders have technical knowledge and authority. Porus, Napolean, Hitler, Ashok, Akbar, Maharana Pratap were themselves great warriors. The greatest example of authority is Lord Shiva of the Hindus. He will reward a sincere and hardworking follower sometimes a flatterer too yet he can destroy each and everybody on earth. Thus leader should be able to give reward and punishment together. It is the authority which is the crux of the leadership, yet to keep leadership sustaining, authority should be used sparingly and with the avowed object to achieve goal or vision of the organization and not due to eccentric decision by the leader. He should have decentralized system of decision-making. Then only he can associate and can have willing support of his men and women in the organization. The CEO of the manufacturing unit should knowhow to operate machines. Similarly, CEO of a software company should knowhow to operate computer like Bill Gates of Microsoft.

He himself is a computer wizard. The CEO of Satyam Ramlinga Raju lost interest in company because he was earlier in some other business and started Satyam Software Company as a matter of convenience for profit-making. The leaders in Infosys were successful because they shared even profit and wealth (shares) of the company with their employees and menial staff. Wipro's success lies in its communication system with the employees in a transparent manner. Even, CEO of Wipro Premji asked his only son to work in some other company and then he allowed his son to join on some unimportant assignment in Wipro. The popularity of Smt. Sonia Gandhi as a leader is due to the fact that she has not accepted the Prime Ministership. Even her heir and son Rahul Gandhi refused to be the Minister time and again in the Union Cabinet. Thus leadership is in 'giving' and not in taking. The organization survives if it has broad objective of public welfare. Fully profit oriented organizations go into oblivion very soon. Social welfare activities of multinationals keep them in good books of public and save them from wrath of public in spite of they keep on making profits. Thus, leader should not only be socially responsible, he should also look to be socially responsive.

Leaders like Pt. J.L. Nehru worked for eighteen hours to see young India to line up with the developing countries. Dhirubhai Ambani's constant struggle with problems and work made him popular with the shareholders of Reliance and they opened out their purse for him. As a result, Reliance has come up at number one slot in India and one of the top fortune 500 companies in the World.

The first Home Minister of India Vallabh Bhai Patel is known as Iron Man of India because he demonstrated unusual energy, vigour and action to merge erstwhile kings, princes and other regional lords with the Union of India after independence. So also Smt. Indira Gandhi is known for her vast reserve of energy. She was never cowdowned by anybody. Even the foreign heads of states used to envy her straightforwardness, self-confidence and energy. Though old leaders and CEOs especially in finance organizations still have final say, yet days are not far off when either they will willingly make way for the young and energetic leaders or the public or workers will throw them out.

It is high time that the leadership genetics should be popularized. Leadership itself is one of the main tools of management. According to one estimate, good leadership can make organization up to 30 per cent without any investment.

It is hoped that proper leadership can make this World a better place to live in.

REFERENCES

Agrawal, Dr. P.K. (1995) : *Motivation and Indian Bureaucracy,* M.D. Publication, New Delhi.

Adair, John (1990) : *Not Bosses but Leaders*, Kogan Limited, 120 Pentonville Road, London, N 19, J.N.

Allio, Robert J. (2003) : *The Seven Faces of Leadership*, Tata McGraw-Hill Publishing Company Ltd., New Delhi.

Bryman, Alan (1992) : *Charisma and Leadership in Organizations*, Sage Publications India Pvt. Ltd., New Delhi.

Chatterjee, Debashis (2003) : *Light the Fire in Your Heart,* Full Circle Publishing, New Delhi.

Choudhury, Dr. N.G. (1988) : *Effective Management*, Orient Longman Ltd., New Delhi.

Davar, Rustoms (1996) : *Creative Leadership*, UBS Publishers & Distributors Ltd., New Delhi.

Heller, Robert (1999) : *Managing People*, Dorling Kindersley Ltd., 9 H. Street, London, WC 2E 8PS.

Sapre, S.A. (1996) : *The Joy of Work*, National Book Trust, India, New Delhi.

Sinha, Jai B.P. (1990) : *Work Culture in the Indian Context*, Sage Publication India Pvt. Ltd., New Delhi.

Bibliography

Adair, John (1990); *Not Bosses but Leaders*, Kogan Limited, 120 Pentonville Road, London, N 19, J.N.

Agarwal, P.K. (1993); "*Land Reforms in India, Constitution and Legal Approach*", M.D. Publications Pvt. Ltd. New Delhi,.

Agarwal, P.N. (1978); "*India's Exports Strategy*", Vikas Publishing House Private Ltd., New Delhi.

Agrawal, Dr.P.K. (1995); "*Motivation and Indian Bureaucracy*", M.D. Publications Pvt. Ltd. New Delhi,.

Agrawal, P.K.(1996); "*India's move towards sustainable development*", M.D. Publications Pvt. Ltd., New Delhi.

AIR 1967 SC p. 104-105

AIR 1987 SC 374

AIR 1988 SC 1037.

AIR 1988 SC 2187

Akyuz, Yilmaz (2008); "*The Global Financial Crisis and Developing Countries*", Resurgence, December, Penang, Third World Network.

Allio, Robert J. (2003); *The Seven Faces of Leadership*, Tata Mc Graw-Hill Publishing Company Ltd. New Delhi,

Anand A.S.; Judicial Review – Judicial Activism – Need for caution p.158, published in *Journal of the Indian Law Institute*, 42, JILI (2000), Bhagwandas Road, New Delhi.

Annual Report (Various issues); Ministry of External Affairs, Government of India, New Delhi.

Arnold, David D. (ed); Market-I, New Delhi.

Aslam, Mehmood ; *Population Development and Food Security, Resource Conservation and Food Security*, Vol. II, Concept Publishing Co., New Delhi.

Athukorala, P. and Sen, K. (2002); "*Saving, Investment and Growth in India*", Oxford University Press, New Delhi.

Aziz, Abdul; *Decentralized governance in Asian countries*, Sage Publications, India Pvt. Ltd., New Delhi.

Batra, Amita and Khan, Zeba (August 2005); "*Revealed Comparative Advantage:*

An Analysis for India and China" Working Paper No. 168, Indian Council for Research on International Economic Relation (ICRIER).

Bende Anthony, Nabende; "*Globalization, FDI, Regional Integration and Sustainable Development*", Ashgate Publishing Company, Burlington, USA.

Bentham, (1979); *Principles of Morals & Legislation* (ed) 324

Bhagwati, J.N. & Padma Desai (1970); "*India: Planning for Industrialization and Trade Policies since 1951*", Oxford University Press, London.

Bhat, T.P.; Guha, Atulan and Paul, Mahua (2006); "*India and China in WTO-Building Complementarities and Competitiveness in External Sector*" Report of Planning Commission, Govt. of India, ISID, New Delhi.

Bhatt, R.K.(1999); "India's Trade Relations with the Middle East", *Saraswati Prakashan Mandir, Allahabad.*

Bhattacharya, Swapan K. and Bhattacharya, Biswa N.(2006); "Free Trade Agreement between China and India: likely Impact and its Implication to Asian Economic Community" ADB Institute Discussion Paper No. 59.

Bhattacharya, Swapan K. and Bhattacharya, Biswa N. (April 2007); "*Gains and Losses of India-China Trade Cooperation- A Gravity Model Impact Analysis*" CES_{IFO} Working Paper No. 1970, Category 7: Trade Policy.

Bryman Alan(1992); *Charisma & Leadership in Organizations*, Sage Publications, New Delhi

Bussiere, Mathieu and Mehl, Arnaud (January 2008); "*China's and India's Role in Global Trade and Finance – Twin Titans for the New Millennium*" Occasional Paper Series No. 80, European Central Bank.

Central Statistical Organization, Government of India.

Chakraborty, Bimal (1997); '*The United Nations and the Third World – Shifting paradigm,*' p. 266 Tata McGraw Hills Publishing Company Ltd., New Delhi

Chakraborty, Bimal (1997); "*The United Nations and the Third World, Shifting paradigms*", Tata Mc Graw-Hill Publishing Co. Ltd., New Delhi

Chandrasekhar, C.P. and Ghosh, Jayati (2004); "*The Market that Failed: Neoliberal Economic Reforms in India*", Left World Books, New Delhi.

Chatterjee, Debashis (2003); *Light the Fire in Your Heart*, Full Circle Publishing, New Delhi.

Chooudhury, Dr. N.G. (1988); *Effective Management*, Orient Longman Ltd., New Delhi,

Davar, Rustoms (1996); *Creative Leadership*, UBS Publishers Distributors Ltd. New Delhi.

Debroy, Bibek (1991); "*China's External Trade Since the Reforms*" Commonwealth Publication, New Delhi.

Desai, Ashok A. (2000); Constitutional Accountability towards Environment' in *Journal of the Indian Law Institute*, New Delhi.

Dimaranan, Betina; Ianchovichina, Elena and Martin, Will (August 2007); "China, India and the Future of the World Economy: Fierce Competition or Shared Growth?" Policy Research Working Paper – 4304, The World Bank Development Research Group Trade team.

Dzever, Sam and Jaussad, Jacques (1999); "*China and India Economic Performance and Business Strategies of Firms in Mid-1990s*.

Economic Survey (Various issues); Ministry of Finance, Government of India, New Delhi.

Ganguly, J.B. (1976); '*A Strategy for Sustainable Development of India* p. 180, in Sustainable Development Strategy ed. By S.P.Shukla & Nandeswar Sharma, Mittal Publications, New Delhi,

Guha, Ashok (1990); "*Economic Liberalisation, Industrial Structure and Growth in India*", Oxford University Press, New Delhi.

Gupta, M.C. & Tiwari, R.K.(ed) (1998); *Restructuring Government*, Indian Institute of Public Administration, Indraprastha Estate, New Delhi.

Hamstad, Tim & Lokesh S.B. (2002); *Allocating Homestead plots as land reforms : Analysis from West Bengal*, Rural Development Institute, USA.

Heller, Robert (1999); *Managing People*, Dorling Kindersley Ltd, 9 H.Street, London, WC 2E 8PS

I.C. Goloknath & Ors. Vs. State of Punjab & another; AIR 1967 S.C. 1643.

International Monetary Fund (2008a); "*Global Financial Stability Report*", Washington D.C. October.

International Monetary Fund (2008b); "*World Economic Out Look*", Washington D.C. October.

Jain, S.P. & Hochgesang Thomes W. (ed) (1995); *Emerging Trends in Panchayati Raj (Rural Local Self Government) in India*, National Institute of Rural Development Rajendra Nagar, Hyderabad.

Justice J.S. Verma (1997); *Equality, Liberty and Prosperity under the Indian Constitution*, Eastern Law House, Calcutta,

Kashyap, Subhas C. (2003); "*The Citizen & Judicial Reforms (Under Indian Polity)*", (ed) Universal Law Publishing Co. (P) Ltd, Delhi (under the auspicious of Centre for Policy Research, New Delhi).

Kelsen, (1934); *Reine Rechtslehre*, chap.8

Kesavananda Bharati Vs. State of Kerala & another 1973–74 SC Cases, para 1947

Mahajan, V.D. (1986); "*Select Modern Governments*", S. Chand & Co. Ltd., Ramnagar, New Delhi

Mahapragya, Acharya (2002); *Mahavir Kaa Arthashastra*, Adarsh Sachitya Sangh MD Publications (P) Ltd., New Delhi.

Middle East Economic Digest (Report)

Mohanty, Manoranjan; CPC's Fourth Generation Ideology, *EPW*, April, 2003.

Monthly Economic Report; Ministry of Finance, Government of India.

Monthly Statistics of Foreign Trade of India, Vol. I & II (Various issues); Director General of Commercial Intelligence & Statistics (DGCI & S), Ministry of Commerce, Government of India, Kolkata.

Monthly Statistics of Foreign Trade of India, Vol. I & II (Various issues); Director

General of Commercial Intelligence & Statistics (DGCI & S); Ministry of Commerce, Government of India, Kolkata.

Naidu, N.Chandrababu & Seventi Nivan (2000); *Plain speaking*, Viking by Penguin Books India (P) Ltd., New Delhi

Narula, Subhash (1988); "*India's Gulf Exports*", Anupana Publications, New Delhi.

Narula, Subhash (1988); "*Gulf Economies in Indian Perspective*", Commonwealth Publications, New Delhi.

National Law School of India (2000); *Handbook on Human Rights for Judicial Officers*, no. 48 p. 5133.

Panagariya, Arvind (2008); "*India: The Emerging Giant*", Oxford University Press, New York.

Panagariya, Arvind; "*India and China: Trade and Foreign Investment*" Columbia University, New York.

Panchayati Raj Update; Ed. Rajkishore for Institute of Social Sciences, 8, Nelson Mandela Road, New Delhi.

Pandey G.P. (ed); G-19, Vijay Chowk, Laxmi Nagar, New Delhi.

Parthasarathy, A. (2008); *Bhagwad Gita*, 1A Landsend, Mumbai.

Patnaik, Prabhat (2007); "Financial Crisis, Reserve Accumulation and Capital Flows", *Economic and Political Weekly*, December 15.

Patnaik, Prabhat (2008); "*The Value of Money*", Tulika Books, New Delhi.

Pound Roscoe, (1958); *The Ideal Element in Law*, Tagore Law Lectures, 1948, Published by University of Calcutta.

Puneet, Pabby (2005); *Global Environment Change & International Law in Environment Development* ed. Jabir Singh, I.K. International Pvt. Ltd., New Delhi.

Reddy, P.L. Sanjeev; *Issues and Themes in Indian Administration*, Indian & Institute of Public Administration, New Delhi.

Report of the India–China Joint Study Group on Comprehensive Trade and Economic Cooperation (2005), New Delhi, India.

Reserve Bank of India (2008); "*Annual Policy Statement for the Year 2008-09*", April.

Reserve Bank of India; "*Weekly Statistical Supplement*", Reserve Bank of India, Mumbai.

Sapre, S.A. (1996); *The Joy of Work*, National Book Trust, India, New Delhi.

Saxena, J.K. (2006); *Swami Vivekananda*, Shanti Prakashan, Bank House, Nainital

Security and Exchange Board of India (SEBI). Mumbai.

Sidhu, W.P.Singh & Yuan, Jing-dong (2005); "*India and China – Cooperation or Conflict* ?" India Research Press, New Delhi, India.

Singh, J.L.; '*50 years of Panchayati Raj & Rural Development*,' Manak Publications Pvt. Ltd. New Delhi.

Singh, S.S. Mishra, Suresh & Pratap Sanjay (1997); *Legislative Status of Panchayati Raj in India*, Indian Institute of Public Administration, New Delhi.

Sinha, Jai B.P. (1980); *Work Culture in the Indian context*, Sage Publication India Pvt. Ltd., New Delhi.

Sinha, Jain B.P. (1990); *Work Culture in the Indian Context*, Sage Publication India Pvt. Ltd. New Delhi.

Stiglitz, Joseph; "*Globalization and its discontents*".

Subha, D.M. (2004); *Bhagad Gita for all times*, Kabir Peace Mission, 1/14 Viswas Khand, Gomti Nagar, Lucknow.

Telegraph; English Daily Publisher for Kolkata.

The Economic Intelligence Unit (EJU); London, UK, Country Report, United Arab Emirates.

Tiwan, Dr. A. K. & Nair, Jeevan (2009); *Food Security and Global Economy*, Pentagon Press, New Delhi.

Toynbee, Dr. Arnold (1976); *We the people* by N.A. Palkhiwala, Strand Book Stall, Mumbai, p-73

Verma, S.K.ed. (2000); *Journal of the Indian Law Institute*, Vol. 42, No. 2-4, April-Dec. 2000, Bhagwan Das Road, New Delhi.

Wadhwani, M & Mishra, S.N. (ed) (1996); *Dreams and Realities Expectations from Panchayati Raj*, Indian Institute of Administration, New Delhi.

Weerahewa, Jeevika and Meilke, Karl (2007); '*Indo-China Trade Relationship: Implications for South Asian Economies*", Paper presentation at the IATRC meeting in Beijing, China.

World Bank, Idea (1997); *Achievements and Challenges in Reading Poverty*, A World Bank, Century, Study, pp. XIII–XIV.

Index